SpringerBriefs in Materials

The SpringerBriefs Series in Materials presents highly relevant, concise monographs on a wide range of topics covering fundamental advances and new applications in the field. Areas of interest include topical information on innovative, structural and functional materials and composites as well as fundamental principles, physical properties, materials theory and design. **Indexed in Scopus (2022).**

SpringerBriefs present succinct summaries of cutting-edge research and practical applications across a wide spectrum of fields. Featuring compact volumes of 50 to 125 pages, the series covers a range of content from professional to academic. Typical topics might include

- A timely report of state-of-the art analytical techniques
- A bridge between new research results, as published in journal articles, and a contextual literature review
- A snapshot of a hot or emerging topic
- An in-depth case study or clinical example
- A presentation of core concepts that students must understand in order to make independent contributions

Briefs are characterized by fast, global electronic dissemination, standard publishing contracts, standardized manuscript preparation and formatting guidelines, and expedited production schedules.

Charu Lata Dube

Microwave Assisted Synthesis of Materials

Springer

Charu Lata Dube
Central University of Gujarat
Vadodara, Gujarat, India

ISSN 2192-1091 ISSN 2192-1105 (electronic)
SpringerBriefs in Materials
ISBN 978-3-032-01438-2 ISBN 978-3-032-01439-9 (eBook)
https://doi.org/10.1007/978-3-032-01439-9

This Springer imprint is published by the registered company Springer Nature Switzerland AG
The registered company address is: Gewerbestrasse 11, 6330 Cham, Switzerland

Preface

The microwave processing of materials is a rapid, energy efficient, and environmentally friendly technique for sintering/calcining different materials. Microwave energy has been in use for over 90 years in a variety of applications, such as communications, food processing, rubber vulcanization, textile/wood products, and drying of ceramic powders. It has now emerged as an important technique for materials processing in numerous industrial applications. It has been revealed by researchers that microwave-assisted sintering can lead to a reduction of 86% in sintering time and energy savings of 96% in comparison to conventional sintering methods. There are two options of microwave (MW) processing: multimode and single mode. In multimode processing, the electromagnetic energy is distributed over several modes (through E- and H-fields) and over a wide volume. In single mode, almost the entire energy is confined to a single mode (either in E- or H-field) and within a smaller volume, which can result in a more efficient interaction with certain materials. Microwave techniques are capable of processing almost all types of inorganic/organic materials. Besides, it is known to result in phase transformation from polycrystalline to amorphous/nanocrystalline phase and alloy formation from elemental constituents in a very short time and even in different morphologies. There are books published on microwave processing of oxides, electroceramics, metals, etc. However, there has not been an adequate account of MW processing of carbon-based composites, functionally graded materials, glass ceramics, dilute magnetic semiconductors, and functional materials for biomedical applications. This book is aimed at filling this gap.

In this book, the microwave processing of the technologically relevant materials such as graphene oxides, metal oxides, and important classes of semiconductors (group IV and II–VI compound semiconductors) is described at length. Besides, this book provides in-depth insight into the interaction of microwaves with different types of materials, with a special focus on semiconducting materials.

This book will serve as an introduction to this emerging technique of material processing for post-graduate students and researchers engaged in the area of Materials Science across the globe.

Vadodara, Gujarat, India — Charu Lata Dube

Contents

1 **Introduction to Microwave Processing of Materials** 1

2 **Microwave Processing Equipment and Modes**. 19

3 **Mechanism of Microwave–Material Interaction** 29

4 **Microwave Processing of Semiconducting Materials** 39

5 **Microwave Processing of Dilute Magnetic Semiconductors** 53

6 **Microwave Processing of Technologically Important Materials**. 63

Chapter 1
Introduction to Microwave Processing of Materials

Abstract This chapter broadly introduces microwaves and microwave (MW) processing of different types of materials ranging from metals to insulators. A detailed comparison between the conventional method and MW processing is made in this chapter. A brief account of historical background and recent trends of MW processing is also presented. The characteristics of MW processing in different modes along with their advantages and applications in materials processing are discussed towards the end of this chapter.

1.1 Introduction

A material can be modified in terms of its crystal structure, microstructure, morphology, and composition by suitable processing to obtain desired properties, which make the material technologically useful in various fields and applications. The processing, in general using various techniques, can result in crystallization, controlled grain growth in polycrystalline materials, formation of amorphous/nanostructured materials, epitaxial growth of thin/thick films, alloy formation, and impurity-substituted materials, e.g. electronic and magnetic semiconductors. In most techniques employed for forming a desired phase/microstructure of a solid, the solid is first converted into a liquid or a vapour state by the application of thermal energy, lasers, and/or sputtering which then is suitably condensed to yield the desired structure/morphology. Different processing techniques which are employed to form the above stated morphologies of the materials include solid-state sintering, physical vapour deposition, chemical vapour deposition, molecular beam epitaxy, and sputtering [1]. These are quite complicated, elaborate, expensive, and dedicated to yield only one phase of materials. In contrast, microwave processing is a simple and economic solution for the processing of materials [2].

Microwaves lie between radio waves and infrared frequencies in the electromagnetic spectrum as shown in Fig. 1.1 [3]. Microwaves, consisting of electric (E) and magnetic (H) fields perpendicular to each other, propagate in the direction perpendicular to the plane containing these fields, as shown in Fig. 1.1 [4].

C. L. Dube, *Microwave Assisted Synthesis of Materials*, SpringerBriefs in Materials, https://doi.org/10.1007/978-3-032-01439-9_1

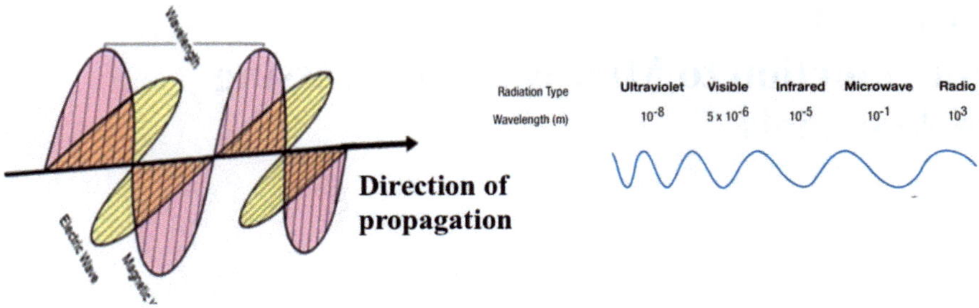

Fig. 1.1 The MW range in the electromagnetic spectrum

The Poynting Eq. (1.1) states that the magnitude of electric field (E_0) is equal to c times the magnitude of magnetic field (H_0).

$$E_0 = cH_0 \tag{1.1}$$

Here E_0: the magnitude of electric field.

H_0: the magnitude of magnetic field.

c: velocity of electromagnetic wave in vacuum (velocity of light).

Microwave (MW) covers a broad frequency range (300 MHz to 300 GHz). The Federal Communications Commission (FCC), however, has allocated two frequency bands of microwaves, centred at 2.45 and 915 MHz, for industrial, scientific, and medical applications. Both the frequencies provide a good compromise between penetration depth (discussed in Chap. 3) in most of the materials and an efficient absorption of microwave by the material. The polar molecules including water, possessing appropriate dielectric constant, when used as a solvent and/or a dispersing medium, couple well at these frequencies. The microwave processing is a relatively new development in processing of materials which has the most profound and wide-ranging impact in satisfying the growing and challenging demands for improved performance of materials and cheaper products/devices in various fields. This is especially true nowadays as both the private sector (industry) and government all over the world in their new policy are focusing on new goals, such as better performance, faster processes, energy savings, and environmentally friendly techniques. The microwave processing of materials fulfils these goals as it is a rapid, green, and energy efficient technique for material synthesis. Besides, the microwave processing has tremendous capabilities, including the formation of non-equilibrium phase of intermetallic compounds [5], decrystallization in ferrites [6], sintering of different materials through precise and controlled volumetric heating [7], recrystallization [8], nanocrystallization of soft-magnetic amorphous alloys [9], rapid crystallization of PZT thin films [10], development of transparent ceramics [11], steel making [12, 13], improved corrosion-resistance steel [14, 15], synthesis of nanoparticles [16–18], and even the sintering of metallic materials [19]. In addition, it is capable of forming alloys beyond equilibrium composition [20], multiferroics [21], metal composites [22], and magnetic semiconductors [23, 24], nanostructural growth [25], and alloys [26]. All this is achieved due to substantial enhancement in reaction and diffusion kinetics during interaction of microwaves with the starting materials [27]. Furthermore, certain products can only be synthesized by either microwave processing or a combination of microwave and conventional heating techniques [28].

Most of the heat in a conventional furnace is absorbed in the first few microns of the exterior layers of a sample, producing *shielding effects* for the interior of the sample that is not immediately subjected to the radiation. Therefore, longer duration is needed by the sample to attain the same temperature. Each section of the sample experiences a different thermal history. The conventional process reduces the product reliability/reproducibility.

1.2 Microwave Processing of Materials

In MW processing, a material absorbs electromagnetic radiation and gets heated. Depending upon the processing temperature, the material can undergo grain growth, phase change, decrystallization, nano-crystallization, etc. The interaction of materials with microwave fields is an intricate process and strongly depends upon the nature of materials. The presence of MW field (E and H) drastically perturbs the movement of charge carriers, orientation of dipoles, alignment of magnetic domains, spinning direction of electrons, molecular vibrations, etc. All these perturbations on atomic scale can lead to different kinds of loss phenomena, such as dielectric, conduction, hysteresis, eddy current, joule heating, etc. finally resulting in the rise in the temperature of the material, leading to a desired morphology/microstructure, etc. The detailed mechanisms of microwave–materials interaction for different materials are discussed in Chap. 3.

Microwave processing has many advantages over conventional methods. Some of these advantages are summarized in Table 1.1 and described in Sect. 1.3.

1.2.1 Historical Background of Microwave Processing

To quote [29]: High-frequency heating really started when engineers working on short-wave transmitters contracted artificial fevers. The great virtues of this kind of heat are as follows: The heat is generated directly in the object itself; no transfer of heat is involved. Associated apparatus need not be heated. The surfaces of the material need not be affected. The people who work with the equipment have cooler working conditions. No gases are involved, and thus, the likelihood of corroded surfaces is eliminated. The material can be heated from the inside-out. Finally, objects of unusual size or shape can be heated.

Microwave energy has been in use for over 90 years in a variety of applications, such as rubber vulcanization, food processing, communications, textile/wood

Table 1.1 Comparison of conventional processing and MW processing

Conventional processing	MW processing
Heat is transferred to the specimen via radiation, conduction, and/or convection	The material is directly heated through its interaction with microwaves
Nonuniform heating which may lead to varied microstructure across a material	Uniform and volumetric heating leading to uniform microstructure across the material
Time consuming (many hrs/days)	Short processing time (generally a few mins), leading to energy saving and hence cost-effectiveness
High sintering temperature and long processing time	Lower temperature requirement
Results in non-stoichiometric compound	Results in stoichiometric compound with improved quality and properties
Uncontrolled heating	Controlled heating
Low heating rates	High heating rates (>400 °C/minute) and hence minimization of grain growth
Established phases	Synthesis of new/special materials and new market potential

products, and drying of ceramic powders [30]. The heating effect of microwaves was fortuitously found by American engineer Percy Spencer, while he was exploring microwaves for radar applications in 1945 [31]. The use of microwaves for materials processing started with low temperature (500 °C) application of MW processing, such as drying, binder burnout, and dewaxing die. The processing ceramic materials with MW energy were very well established by the 1950s and had been probed by several groups on a limited basis during that period. In 1975, Sutton observed that, while investigating MW drying of high alumina castables, MW can heat the ceramics. Using an industrial MW oven lined with microwave transparent refractories, temperatures exceeding 1400 °C were achieved. This investigation demonstrated that MW energy could be used for processing full-scale ceramic products and that MW could be employed for both low- and high-temperature processing [32]. Microwaves are a versatile tool to process different materials, e.g. polymers [33], woods [34], ceramics [35], glass matrix composites [36], metallic powder [37], etc. Besides, the microwave processing can be used for the development of novel structures [38].

1.2.2 Recent Trends in Microwave Processing

The use of MW in the processing of ceramics, mostly the *sintering of materials*, picked momentum only in the late 1970s and continued with great enthusiasm and excitement in the 1980s. The earlier excellent reviews [39–42] have summarized the status of MW processing research till 1996.

During this period, besides the full commercialization of low temperature MW drying and food processing, the MW heating and sintering of traditional and advanced ceramics, composites, alumina, uranium oxides, silica, zeolites, barium strontium titanates, hard glass-ceramic coating, Si_3N_4, hydroxyapatite, NASICON for nuclear waste disposal, etc. were investigated but not commercialized [43–47]. Activated carbon, a high-quality product, has been successfully prepared from coconut shell chars in pilot-scale microwave heating equipment in a microwave furnace at 60 kW (equipped with 60 magnetrons of 1 kW each) operating in an inert atmosphere at 2.45 GHz at a decreased reaction temperature and shorter processing time as compared to conventional heating [48]. Recent reviews [49–51] have summarized the latest developments in the field of MW processing/sintering and synthesis of inorganic solids.

It is known that the aggressive scaling down of semiconducting devices has driven most of the advances in the electronics industry as the reduction in feature size increases the switching speed of the device and enhances the performance. Besides, miniaturization makes it energy efficient and economic. However, there are significant drawbacks, e.g. short-channel effects, threshold voltage roll off, reverse short channel effects, increased gate–source/drain capacitance, and drain-induced barrier lowering [52]. By using MW energy, these problems can be reduced. The MW processing of semiconductors is emerging as potentially beneficial method for semiconductor industry. Thompson *et al.* [53] demonstrated the use of MW energy as a highly effective mechanism for rapid thermal annealing. They have shown the efficient deposition of high-frequency microwave energy into the top several microns of a semiconducting material, which can minimize the hurdles related to high-temperature processing of devices. Recently, the microwaves are being used in material processing for synthesis of biocompatible potential nano-drug carriers, such as superparamagnetic iron oxide nanoparticles [54, 55]. MW energy was employed for reduction of graphene oxide, to obtain graphene-like material in bulk [56]. It is reported that microwaves achieve accelerated catalytic activity and enhanced selectivities in the alkylation of phenols [57]. The MW-assisted fabrication of quantum dots (QDs) represents a growing research field in nanomaterial research. It reproducibly produces large quantity of QDs in a short time. In a recent review article, synthesis of quantum dots (graphene/carbon/semiconducting) for bioapplications has been reviewed [58].

1.3 Characteristics of Microwave Processing

The microwave processing of materials is associated with peculiar characteristics, which are described in the following subsections.

1.3.1 Volumetric Nature

In MW processing due to the penetrative power of MW, heat can be generated from within the materials itself. MW energy is absorbed directly, instead of transferring through conduction and/or convection currents (occurs in conventional heating).

This heating during MW processing is therefore volumetric in nature. The volumetric nature provides high heating rates (>400 °C/minute) and leads to minimization of grain growth. Due to the volumetric nature of MW–material interaction, it is possible to sinter many materials at a much lower temperature and shorter time than required in conventional methods. The use of MW processing reduces typical sintering times by a factor of 10–100 in many cases, thereby making it fast and economical. Further, it is possible to retain the initial fine grain structure without using grain growth inhibitors.

1.3.2 Inverted Temperature Profile

During MW processing, heat is generated first within the volume of the material and then radiates outwards, Consequently, the surface temperature is lower than that of the core temperature. This creates an inverted temperature profile and provides unique MW processing effect. During conventional heating, the surface gets heated up first and is at a higher temperature than the interior; this results in the surface reaction completing before the interior is fully reacted. In some cases, surface pores may close prematurely, preventing mass transport of gas reactants to the centre needed to complete the reaction. In contrast, the microwave processing allows the reactant gases to permeate the specimen and diffuse to the hotter central region until the reaction is completed. Furthermore, it minimizes the surface degradation of the specimen.

1.3.3 Anisothermal Effect

Microwave absorption strongly depends on the material properties. Different materials depending upon their dielectric permittivity/magnetic permeability interact with MW very differently. In a multiphasic system, if one phase is a MW absorber and the other is not, then there will be huge, localized temperature gradient, a situation known as *anisothermal condition* and consequent effect is known as anisothermal effect, which may cause drastic enhancement in the reaction kinetics and material diffusion. In the conventional heating of the same system, the temperature rise will be slow and hardly any reaction/diffusion would occur. Fig. 1.2 depicts temperature of a two-component ceramic system showing an isothermal condition in a conventional furnace and a localized different temperature (anisothermal condition) in the two-component ceramic system in the MW furnace. The good microwave ceramic system and poor microwave absorbing system are shown in yellow and red colour, respectively. The accelerated kinetics, enhanced diffusion, reduced processing time, and reduced activation energy during microwave processing have been reported [59]. Anisothermal effects induced by microwave fields have been reported in Y_2O_3 (poor microwave absorbing)–Fe_3O_4 (highly microwave absorbing)

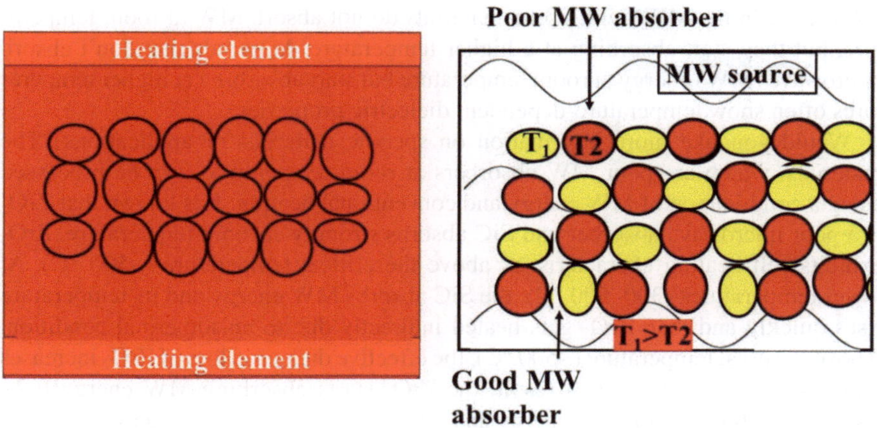

Fig. 1.2 A schematic comparison of heating in conventional and MW processing of a two-ceramic system. Temperature of both the ceramic components is same in the conventional furnace, whereas it is different in the MW furnace

system using solid-state synthesis carried out in a 2-kW tubular microwave furnace operating at a microwave frequency of 2.45 GHz [60]. Using anisothermal situation, garnets, spinels, doped magnetite, and many ferrites have been synthesized and sintered in a time period duration which is 1/100th of time normally required in a conventional process [55, 61].

1.3.4 Self-Heating

The relative static permittivity (dielectric constant) describes the response of a dielectric material to an applied static electric field. The response of a dielectric material to a changing electric field is measured in terms of complex permittivity. The complex permittivity changes with the frequency of the applied electric field and temperature of the specimen. The real part of the permittivity corresponds to the energy stored in the dielectric material, and the imaginary part is linked to various dielectric loss processes. The effective dielectric loss (ε''_{eff}), encompassing different kinds of polarization losses, such as electronic, ionic, dipolar, interfacial, and conduction losses, strongly depends on the temperature. Many ceramics exhibit an abrupt increase in the effective dielectric loss with increasing temperature. The MW power absorbed (P_a) by dielectric materials is proportional to effective dielectric loss. Consequently, there is a corresponding increase in the P_a and the heating rate of the ceramic as the temperature increases. The reason for this abrupt change in ε''_{eff} is most likely due to smoother dipole rotation. The temperature where the abrupt change in ε''_{eff} occurs is referred to as the *critical temperature*, T_c. Below T_c, an external energy source is required to heat ceramic. Above T_c, the materials get *self-heated* in the MW field. Some materials do not absorb MW at room temperature, but they start absorbing at a higher temperature. Materials that don't absorb microwave (MW) energy at room temperature but start absorbing at higher temperatures often show temperature-dependent dielectric properties.

Would you like more information on specific materials or applications? The materials, which are poor MW absorbers at room temperature, can be processed using a combination of MW energy and conventional heating. It is known that ZrO_2 is a poor microwave absorber and SiC absorbs strongly at room temperature. ZrO_2 exhibits self-heating characteristics above the critical temperature (~500 °C). At lower temperatures (300–500 °C), the SiC absorbs MW energy and its temperature rises quickly and then ZrO_2 gets heated indirectly due to anisothermal condition. Above a critical temperature (~500 °C), the effective dielectric loss of ZrO_2 increases with rising temperature; as a result, the ZrO_2 starts absorbing MW energy [62]. Using a combination of MW energy and IR heating, Al and Al/Cu nanocomposites with superior properties had been synthesized [63].

1.3.5 Self-Controlled Heating

As discussed in the previous section, properties of materials change with temperature. Some materials absorb MW at room temperature, but at higher temperature they stop absorbing owing to changes in their properties. Carbonaceous materials are good MW absorber. For example, graphitic/graphene compounds have abundant delocalised π electrons which facilitate absorption of MW and get heated via joule heating [64]. However, in the presence of oxygen containing groups, some of the graphitic compounds (such as graphite flakes and black carbon) get oxidized and graphite oxide is formed. Since the oxide does not have delocalised π electrons, these compounds stop absorbing microwaves. The entire process is an excellent example of self-controlled heating during MW processing.

1.3.6 Non-thermal Microwave Effects

The non-thermal microwave effects open a potential field of reaction mechanisms and kinetics [4]. It is found that MW-synthesized microporous materials (molecular sieves) have higher population of nuclei of smaller size than in the samples synthesized by a conventional method [65]. The cumulative degree of acceleration in case of MW synthesis was found to be many folds as compared to conventional methods. The presence of MW fields (E- and H-field) accelerates the synthesis process. The acceleration occurs due to the distinctive interaction of MW with the irradiated specimen and is called *non-thermal effect* [66]. The observed increase in microwave-assisted reaction is very similar to the result of purely thermal/kinetic effects [67]. The enhancement in number of critical size nuclei was attributed to non-thermal microwave effects. This effect in some cases such as organic compounds (xylene and dibutyl ether) and metal powder can be understood with the help of Arrhenius law [68, 69].

$$k = Ae^{\frac{E_a}{RT}}$$

Here, k is the rate constant; A is the pre-exponential factor; E_a is the activation energy; R is the universal gas constant; and T is the absolute temperature.

In case of organic compounds, it is believed that strong excitation of molecular rotations by microwaves could *weaken* certain bonds by affecting their vibrations [63]. The sintering behaviour of different metal powders was investigated under single-mode microwave radiation (single-mode processing is discussed in Sect. 4.2). It is found that the free energy of the high-temperature phase is lowered by microwave radiation, leading to a reduction of the transformation temperature [64]. It is well understood that the electric component of microwaves is responsible for the polarization of dipolar molecules or intermediates and thereby leads to the change in the pre-exponential factor or the activation energy term in the Arrhenius equation [70]. It is believed that the presence of MW fields (E- and H-field) enhances reaction rate and brings down activation energy [71].

During microwave processing, the unique characteristics of microwaves (penetrating radiation, controllable electric field distributions, rapid acceleration of dielectric losses above a critical temperature, differential absorption, and self-limiting reactions) provide substantial process flexibility to tune the process parameters and hence to get specific materials with desirable properties [72]. Field-induced decrystallization of silicon had been realized owing to microwave non-thermal effect [73]. This makes microwaves a versatile tool for processing wide range of materials starting from insulators to semiconductors [23, 74].

1.4 Modes of Microwave Processing

1.4.1 Multimode Microwave Processing

The multimode MW processing is carried out in a metallic applicator. The domestic MW oven is an example multimode applicator. Typically, multimode MW processing unit consists of MW source, an applicator, rotating plate, and a mode stirrer. The schematic diagram of multimode MW processing unit is shown in Fig. 1.3.

An applicator is designed to maintain a distinctive distribution of electromagnetic field (E- and H-field), known as transverse electromagnetic (TEM) mode. The TEM mode has an attribute to have both the fields (E- and H-field) perpendicular to the direction of propagation of electromagnetic radiation. The different types of modes are discussed in Chap. 2 (Sect. 2.3.2) in detail. In a multimode system, there are over 100 modes present inside the processing unit, but all the modes are mixed,

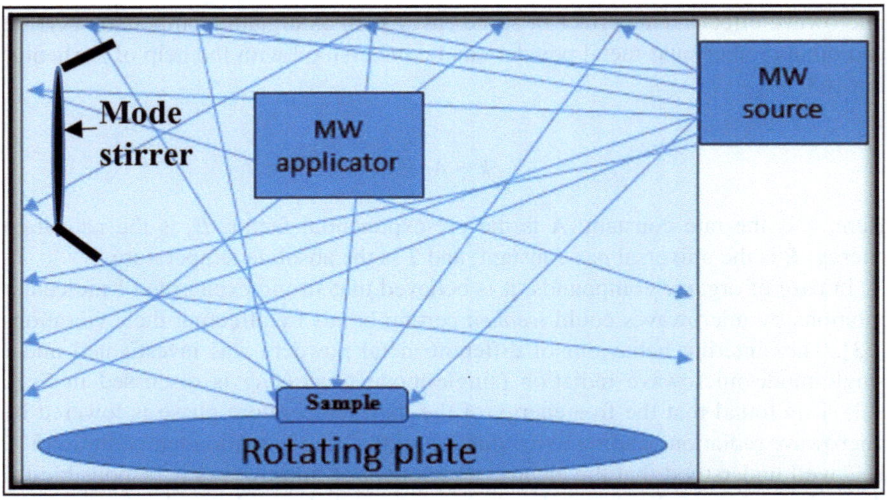

Fig. 1.3 The schematic diagram of multimode microwave processing unit

and one cannot study the effect of individual field/mode on the material processing. In addition, in a multimode system, the MW energy is distributed over all these modes and over the entire volume.

Alloy compacts of 92.5 W–6.4Ni–1.1Fe were sintered in a multimode microwave applicator (2.45 GHz, 2 kW). The MW sintered compacts at 1500 °C for 20 min, exhibited finer grain structure against conventionally sintered samples [75]. The austenitic (316 L) and ferritic (434 L) had also been compacted in multimode MW applicator. It is found that the sintering time in a microwave applicator was reduced by ~90%, and the sintered density was higher using a multimode microwave applicator as compared to conventional furnace [15]. The metallic tungsten powder efficiently couples with microwaves and gets heated very rapidly, consequently resulting in tremendous reduction in the sintering time [76–78]. Microwave heating of porous metal which has particle size less than microwave penetration depth is recently reported in 1999 [19]. Pure copper powder having a varying porosity efficiently couples with microwaves at 2.45 GHz in a multimode microwave furnace [37]. Refractory metals/alloys: W, Mo, and e and their alloys W-Ni-Cu and W-Ni-Fe were sintered in a microwave applicator operating at 2.45 GHz [4, 20].

Field uniformity has, however, been a concern in the multimode applicators, because the wavelengths are several centimetres. The disadvantage of multimode system is the presence of nonuniform field distribution due to constructive interference of the electromagnetic waves propagating inside the system. It results in the rapid increase in dielectric loss (ε'') with temperature ($d\varepsilon''/dT$) at points having higher magnitude of fields. The sample is, therefore, not heated uniformly and localized hot spots come into picture. The development of *hot spots* is usually referred to as *thermal runaway* [79]. The *thermal runaway* could be curtailed by hybrid heating, i.e. heating the sample with *radiant heat* as well. Nonetheless, the monitoring of the input microwave power during the heating process can avoid the *thermal runaway* to some extent [80]. The surface temperature can be used as a control signal in the feedback power control loop, and thus, it is possible to implement any prescribed temperature–time schedule for a desired constant heating rate by varying the MW power. However, it has been found that despite such a control of the surface temperature, the temperature in the core increases uncontrollably for materials having sharp temperature dependence of microwave absorptivity and by transfer of heat from the outer layer to the inner core of the particle through conduction [49].

The internal electric field in a material is controlled by the MW power (P_a) inside the applicator which can vary broadly in the processing units (applicators). Mode stirrers and rotating platforms provide some enhancement in the uniformity of MW field as seen by the sample through time averaging. The MW field uniformity can be made better by changing the dimensions of the applicator. The applicator's dimensions must be much substantially greater than the wavelength of MW (~123 cm @ 2.45GHz). With these modifications, large samples can be processed in a reasonably uniform field in a multimode applicator.

1.4.2 Single-Mode Processing

There are several modes of the MW propagation which can be utilized to our advantage for the microwave-assisted processing of materials. Based on the modes used, the microwave processing can be classified into three types: multimode single transverse electric (TE/H) mode and transverse magnetic (TM/E) mode. In multimode processing, several modes are present inside the processing unit (as discussed above in Sect. 4.1).

In the transverse electric (TE/H) mode, the component of the electric field/magnetic field in the direction of propagation of MW is zero/nominal [81]. Conversely, in the transverse magnetic (TM/E), the component of the magnetic field in the direction of propagation of MW is zero/nominal, as shown in Fig. 1.4. At position 1 and position 2 (Fig. 1.4), the amplitude of H-field is maximum and minimum, respectively. Therefore, in the same applicator both the maximum and minimum H-field can be obtained for the materials processing. Besides, both the maximum and minimum E-field can also be obtained in the same applicator for the materials processing.

The TE_{mnl}/TM_{mnl} modes are present in rectangular and cylindrical waveguides. The indices m, n, and l are the order of the modes, and they can be any integer, 0, 1, 2, etc. Depending upon the size of the waveguide, different modes exist in a waveguide. Each TE and TM wave in a waveguide has different field configurations [20]. The field configurations for TE/TM modes are discussed in detail in Chap. 2.

Single-mode applicators/cavities are designed and employed to minimize unnecessary distribution of energy and formation of hot spots during MW processing. In a single-mode applicator almost, entire energy is confined either in high E- or H-field, resulting in more efficient interaction with certain materials. Besides, there

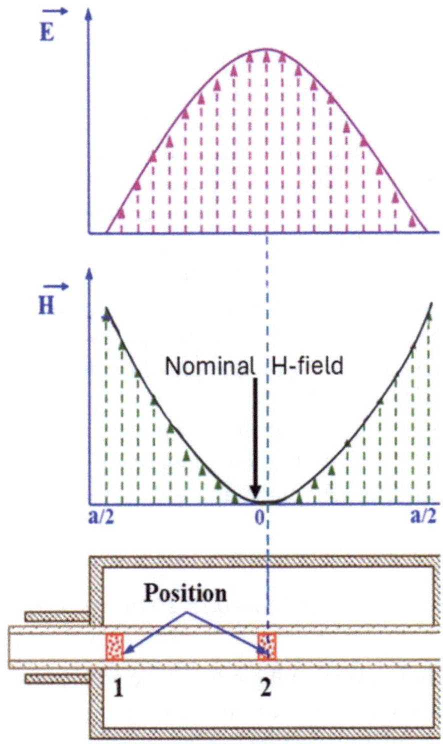

Fig. 1.4 The component of the magnetic field in the direction of propagation of MW is zero/nominal in TM mode in a cylindrical resonator of length *a*

is an additional flexibility of being able to optimize the geometry of the cavity to fit the parameters of a specimen [82], as depending on the dimensions of an applicator, the field distribution inside the cavity can be manipulated.

The E- and H-field of an electromagnetic wave can be isolated [83]. The single-mode technique has been exploited for joining the ceramics [84] by focusing the microwave energy at the joint interface without heating the bulk of the ceramic. Furthermore, the different materials exhibit greatly different heating behaviour in E- and H-field [10]. It is worth mentioning that bulk metals do reflect MWs. Nevertheless, it is found that a powdered metal-compact heats well in an H-field, while in the E-field it hardly reaches to high temperature [56]. Conversely, oxides such as iron oxide, ZnO, and NiO show completely different behaviour and get heated up well in the E-field [6]. Similarly, 70% Si silicon along with 30% Ge has been formed by rapid heating in a pure E-field in a single-mode 2.45GHz microwave cavity. SiGe alloys have exhibited drastically different results in E- and H-fields in TE $_{103}$ cavity. The pure H-field microwave-heated samples produce only single-phase alloys. The MW processing of these materials in E-field has resulted in decrystallization. Furthermore, the H-field processing of E-field synthesized amorphous SiGe sample transforms it into the crystalline phase [85]. The H-field processing of a mixture containing silicon and germanium powder (in different ratios) has resulted in the formation of polycrystalline SiGe alloys within few minutes [86, 87]. Recently, it is reported that E-field processing of polycrystalline SiGe alloy (duly synthesized by using H-field) has resulted in the formation of amorphous SiGe alloys [68].

Diverse kinds of metals such as Co, Fe, and Cu showed a unique heating profile under microwave irradiation. The unique heating profile establishes that the magnetic field interaction contributes largely to heating of the metals [88]. Furthermore, it is reported that magnetic metals heated more successfully when placed within a strong magnetic field, and others heated more beneficially when placed in regions with stronger electric field [3, 89].

Single-mode processing and anisothermal heating both have been successfully employed to prepare doped semiconductors. An anisothermal condition for heating ZnO-Co powder in either H- or E-field had been utilized. In a pure H-field, initially Co powder gets heated due to high magnetic loss, while ZnO powder does not absorb MW power. After long (10 min) exposure to MW H-field, the ZnO-Co powder gets heated to higher temperatures due to conductive heat transfer. In a pure E field, the temperature profile is reversed and ZnO powder has higher temperature causing the heat transfer from ZnO to Co. Thus, anisothermal condition at microscopic level leads to heating of ZnO-Co powder in both H- and E-fields [88].

1.5 Conclusions

This chapter provides an introduction and historical background of microwave (MW) processing, which is a green, rapid, and economic technique. This chapter briefs recent trends and the characteristics of microwave processing, besides its capabilities to process all kinds of materials. The materials processed employing this technique include transparent ceramics, multiferroics metal composites, and magnetic semiconductors. The two modes of microwave processing are introduced. The single-mode MW processing due to much more controlled high field (E or H) distribution results in the faster and energy efficient heating of certain materials to yield desired morphology and microstructure.

References

1. Schütz, M. B., Xiao, L., Lehnen, T., Fischer, T., & Mathur, S. (2018). Microwave-assisted synthesis of nanocrystalline binary and ternary metal oxides. *International Materials Reviews, 63*(6), 341–374.
2. Gurram, N. R., & Nath, T. K. (2018). Investigation of Superparamagnetism in microwave and conventional processed Mn 0.5 Zn 0.5 Fe 2 O 4 nanoparticles. *Journal of Superconductivity and Novel Magnetism, 31*, 815–820.
3. Griffiths, D. J. (2023). *Introduction to electrodynamics*. Cambridge University Press.

4. Nee, A. Y. C. (2014). *Handbook of manufacturing engineering and technology*. Springer. Incorporated.
5. Veronesi, P., Rosa, R., Colombini, E., Leonelli, C., Poli, G., & Casagrande, A. (2010). Microwave assisted combustion synthesis of non-equilibrium intermetallic compounds. *Journal of Microwave Power and Electromagnetic Energy, 44*(1), 45–56.
6. Roy, R., Peelamedu, R., Hurtt, L., Cheng, J., & Agrawal, D. (2002). Definitive experimental evidence for Microwave Effects: radically new effects of separated E and H fields, such as decrystallization of oxides in seconds. *Materials Research Innovations, 6*(3), 128–140.
7. Raihanuzzaman, R., Chuan, L., Xie, Z., & Ghomashchi, R. (2015). 'Microwave Sintering and Its Application on Cemented Carbides'. World Academy of Science, Engineering and Technology, Open Science Index 10. *International Journal of Materials and Metallurgical Engineering, 9*(8), 993–996.
8. Sorescu, M., Diamandescu, L., Ramesh, P. D., Roy, R., Daly, A., & Bruno, Z. (2007). Evidence for microwave-induced recrystallization in NiZn ferrites. *Materials Chemistry and Physics, 101*(2–3), 410–414.
9. Nicula, R., Stir, M., Ishizaki, K., Català-Civera, J. M., & Vaucher, S. (2009). Rapid nano-crystallization of soft-magnetic amorphous alloys using microwave induction heating. *Scripta Materialia, 60*(2), 120–123.
10. Cao, Z., Wang, Z., Yoshikawa, N., & Taniguchi, S. (2008). Microwave heating origination and rapid crystallization of PZT thin films in separated H field. *Journal of Physics D: Applied Physics, 41*(9), 092003.
11. Cheng, J., Agrawal, D., & Zhang, Y. (2000). Fabricating Transparent Ceramics by Microwave Sintering. *American Ceramic Society Bulletin, 79*(9), 71–74.
12. Nagata, K., Ishizaki, K., Kanazawa, M., Hayashi, T., Sato, M., Matsubara, A., & Roy, R. (2007, January). A concept of microwave furnace for steel making in industry scale. In *11th International Conference on Microwave and High Frequency Heating, AMPERE 2007* (pp. 87–90). University of Oradea.
13. Nagata, K., Hara, K., & Sato, M. (2019). Continuous Process of Pig Ironmaking using Focused Microwave Beams at 2.45 GHz. *ISIJ International, 59*(6), 1033–1040.
14. Padmavathi, C., Upadhyaya, A., & Agrawal, D. (2007). Corrosion behavior of microwave-sintered austenitic stainless steel composites. *Scripta Materialia, 57*(7), 651–654.
15. Panda, S. S., Singh, V., Upadhyaya, A., & Agrawal, D. (2006). Sintering response of austenitic (316L) and ferritic (434L) stainless steel consolidated in conventional and microwave furnaces. *Scripta Materialia, 54*(12), 2179–2183.
16. Bondioli, F., Bonamartini Corradi, A., Ferrari, A. M., & Leonelli, C. (2008). Synthesis of Zirconia Nanoparticles in a Continuous-Flow Microwave Reactor. *Journal of the American Ceramic Society, 91*(11), 3746–3748.
17. Mangalaraja, R. V., Mouzon, J., Hedström, P., Camurri, C. P., Ananthakumar, S., & Odén, M. (2009). Microwave assisted combustion synthesis of nanocrystalline yttria and its powder characteristics. *Powder Technology, 191*(3), 309–314.
18. Pires, F. I., Joanni, E., Savu, R., Zaghete, M. A., Longo, E., & Varela, J. A. (2008). Microwave-assisted hydrothermal synthesis of nanocrystalline SnO powders. *Materials Letters, 62*(2), 239–242.
19. Roy, R., Agrawal, D., Cheng, J., & Gedevanishvili, S. (1999). Full sintering of powdered-metal bodies in a microwave field. *Nature, 399*(6737), 668–670.
20. Mondal, A., Agrawal, D., & Upadhyaya, A. (2010). Microwave sintering of refractory metals/alloys: W, Mo, Re, W-Cu, W-Ni-Cu and W-Ni-Fe alloys. *Journal of Microwave Power and Electromagnetic Energy, 44*(1), 28–44.
21. Prado-Gonjal, J., Villafuerte-Castrejón, M. E., Fuentes, L., & Moran, E. (2009). Microwave–hydrothermal synthesis of the multiferroic $BiFeO_3$. *Materials Research Bulletin, 44*(8), 1734–1737.
22. Gupta, M., Leong, W. W., & Wong, W. L. (2007). *Microwaves and Metals*. John Wiley and sons.

23. Subasri, R., Asha, M., Hembram, K., Rao, G. V. N., & Rao, T. N. (2009). Microwave sintering of doped nanocrystalline ZnO and characterization for varistor applications. *Materials Chemistry and Physics, 115*(2–3), 677–684.
24. Dube, C. L., Kashyap, S. C., Dube, D. C., & Agarwal, D. K. (2008, November). Microwave processing of ZnO based dilute magnetic semiconductors. In *2008 International conference on recent advances in microwave theory and applications* (pp. 126–129). IEEE.
25. Dube, C. L., Kashyap, S. C., Dube, D. C., & Agarwal, D. K. (2009). Synthesis of silicon nanowires using microwave energy for optical devices. *International Journal of Green Nanotechnology: Materials Science & Engineering, 1*(1), M11–M15.
26. Dube, C. L., Kashyap, S. C., Dube, D. C., & Agrawal, D. K. (2013). Rapid alloying of tetrahedral diamagnetic semiconductors in microwave H field. *Journal of Alloys and Compounds, 571*, 75–78.
27. Dube, D. C., Ramesh, P. D., Cheng, J., Lanagan, M. T., Agrawal, D., & Roy, R. (2004). Experimental evidence of redistribution of fields during processing in a high-power microwave cavity. *Applied Physics Letters, 85*(16), 3632–3634.
28. Kashyap, S. C. (2009, December). Microwave processing-A new dimension in synthesis of materials. In *In 2009 Applied Electromagnetics Conference (AEMC)* (pp. 1–4). IEEE.
29. National Research Council, Division on Engineering, Physical Sciences, National Materials Advisory Board, Commission on Engineering, Technical Systems, & An Emerging Industrial Technology. (1994). *Microwave processing of materials* (Vol. 473). National Academies Press.
30. Singh, K., Sharma, N., Gurram, J. R., & Gopalan, S. (2013). U.S. Patent No. 8,344,301. Washington, DC: U.S. Patent and Trademark Office.
31. Spencer, P. L. (1950). U.S. Patent No. 2,495,429. Washington, DC: U.S. Patent and Trademark Office.
32. Yoshikawa, N., Ishizuka, E., Mashiko, K., Chen, Y., & Taniguchi, S. (2007). Brief review on microwave (MW) heating, its application to iron & steel industry and to the relevant environmental techniques. *ISIJ International, 47*(4), 523–527.
33. Lewis, D. A. (1992). Microwave processing of polymers-an overview. *MRS Online Proceedings Library Archive, 269*, 21.
34. Vongpradubchai, S., & Rattanadecho, P. (2009). The microwave processing of wood using a continuous microwave belt drier. *Chemical Engineering and Processing: Process Intensification, 48*(5), 997–1003.
35. Fliflet, A. W., Bruce, R. W., Kinkead, A. K., Fischer, R. P., Lewis, D., Rayne, R., & Schoen, P. E. (1996). Application of microwave heating to ceramic processing: Design and initial operation of a 2.45-GHz single-mode furnace. *IEEE Transactions on Plasma Science, 24*(3), 1041–1049.
36. Boccaccini, A. R., Veronesi, P., & Leonelli, C. (2001). Microwave processing of glass matrix composites containing controlled isolated porosity. *Journal of the European Ceramic Society, 21*(8), 1073–1080.
37. Mondal, A., Agrawal, D., & Upadhyaya, A. (2008). Microwave heating of pure copper powder with varying particle size and porosity. *Journal of Microwave Power and Electromagnetic Energy, 43*(1), 5–10.
38. Hawley, M., Wei, J., & Adegbite, V. (1994). Microwave processing of polymer composites. *MRS Proceedings, 347*, 669.
39. Clark, D. E., & Sutton, W. H. (1996). Microwave processing of materials. *Annual Review of Materials Science, 26*(1), 299–331.
40. Schiffmann, R. F. (1995). *Commercializing microwave systems: paths to success or failure (No. CONF-9505249-)*. American Ceramic Society.
41. Katz, J. D. (1992). Microwave sintering of ceramics. *Annual Review of Materials Science, 22*(1), 153–170.
42. Sutton, W. H. (1989). Microwave processing of ceramic materials. *American Ceramic Society Bulletin, 68*(2), 376–386.

43. Agrawal, D. (2006, August). Microwave sintering, brazing and melting of metallic materials. In *Sohn International Symposium; Advanced processing of metals and materials Volume 4: New, improved and existing technologies: Non-ferrous materials extraction and processing* (Vol. 4, pp. 183–192).
44. Cheng, J., Agrawal, D., Zhang, Y., & Roy, R. (2002). Microwave sintering of transparent alumina. *Materials Letters, 56*(4), 587–592.
45. Thakur, O. P., Prakash, C., & Agrawal, D. K. (2002). Microwave synthesis and sintering of Ba0. 95Sr0. 05TiO$_3$. *Materials Letters, 56*(6), 970–973.
46. Thakur, O. P., Prakash, C., & Agrawal, D. K. (2002). Dielectric behavior of Ba0. 95Sr0. 05TiO$_3$ ceramics sintered by microwave. *Materials Science and Engineering: B, 96*(3), 221–225.
47. Das, S., Mukhopadhyay, A. K., Datta, S., Das, G. C., & Basu, D. (2008). Hard glass-ceramic coating by microwave processing. *Journal of the European Ceramic Society, 28*(4), 729–738.
48. Vaidhyanathan, B., & Rao, K. J. (1996). Rapid microwave assisted synthesis of hydroxyapatite. *Bulletin of Materials Science, 19*(6), 1163–1165.
49. Hossain, F., Turner, J. V., Wilson, R., Chen, L., de Looze, G., Kingman, S. W., & Dimitrakis, G. (2024). State-of-the-art in microwave processing of metals, metal powders and alloys. *Renewable and Sustainable Energy Reviews, 202*, 114650.
50. Agrawal, D. (2006). Microwave sintering of ceramics, composites and metallic materials, and melting of glasses. *Transactions of the Indian Ceramic Society, 65*(3), 129–144.
51. Rao, K. J., Vaidhyanathan, B., Ganguli, M., & Ramakrishnan, P. A. (1999). Synthesis of inorganic solids using microwaves. *Chemistry of Materials, 11*(4), 882–895.
52. Neamen, D. A. (2012). *Semiconductor physics and devices: basic principles*. McGraw-Hill.
53. Thompson, K., Booske, J. H., Ives, R. L., Lohr, J., Gorelov, Y., & Kajiwara, K. (2005). Millisecond microwave annealing: Driving microelectronics nano. *Journal of Vacuum Science & Technology B: Microelectronics and Nanometer Structures Processing, Measurement, and Phenomena, 23*(3), 970–978.
54. Saxena, N., Dholia, N., Akkireddy, S., Singh, A., Yadav, U. C., & Dube, C. L. (2020). Efficient microwave synthesis, functionalisation and biocompatibility studies of SPION based potential nano-drug carriers. *Applied Nanoscience, 10*(2), 649–660.
55. Saxena, N., Agraval, H., Barick, K. C., Ray, D., Aswal, V. K., Singh, A., & Dube, C. L. (2020). Thermal and microwave synthesized SPIONs: Energy effects on the efficiency of nano drug carriers. *Materials Science and Engineering, 111*, 110792.
56. Jakhar, R., Yap, J. E., & Joshi, R. (2020). *Microwave reduction of graphene oxide*. Carbon.
57. Yadav, G. D., & Bisht, P. M. (2004). Novelties of microwave assisted liquid–liquid phase transfer catalysis in enhancement of rates and selectivities in alkylation of phenols under mild conditions. *Catalysis Communications, 5*(5), 259–263.
58. Singh, R. K., Kumar, R., Singh, D. P., Savu, R., & Moshkalev, S. A. (2019). Progress in microwave-assisted synthesis of quantum dots (graphene/carbon/semiconducting) for bioapplications: a review. *Materials Today Chemistry, 12*, 282–314.
59. Jacob, J., Chia, L. H. L., & Boey, F. Y. C. (1995). Thermal and non-thermal interactions of microwave radiation with materials. *Journal of Materials Science, 30*, 5321–5327.
60. Peelamedu, R. D., Roy, R., & Agrawal, D. K. (2001). Reaction kinetics and anisothermal effects in microwave fields: System Y_2O_3–Fe_3O_4. *Journal of Materials Research, 16*(10), 2770–2772.
61. Peelamedu, R. D., Roy, R., & Agrawal, D. (2001). Anisothermal reaction synthesis of garnets, ferrites, and spinels in microwave field. *Materials Research Bulletin, 36*(15), 2723–2739.
62. Sutton, W. H. (1993). Microwave processing of materials. *MRS Bulletin, 18*(11), 22–29.
63. Nawathe, S., Wong, W. L. E., & Gupta, M. (2009). Using microwaves to synthesize pure aluminum and metastable Al/Cu nanocomposites with superior properties. *Journal of Materials Processing Technology, 209*(10), 4890–4895.
64. Dubey, A., Bhavsar, N., Pachchigar, V., Saini, M., Ranjan, M., & Dube, C. L. (2022). Microwave assisted ultrafast synthesis of graphene oxide based magnetic nano composite for environmental remediation. *Ceramics International, 48*(4), 4821–4828.

65. Jhung, S. H., Jin, T., Hwang, Y. K., & Chang, J. S. (2007). Microwave effect in the fast synthesis of microporous materials: which stage between nucleation and crystal growth is accelerated by microwave irradiation? *Chemistry–A European Journal, 13*(16), 4410–4417.
66. Rodrigues, B. S., Vicente, M. R., & Souza, J. S. (2023). Investigating the role of microwave thermal and non-thermal effects on WO 3-graphene oxide composite synthesis. *RSC Advances, 13*(38), 26794–26803.
67. Rodríguez, A. M., Prieto, P., de la Hoz, A., Díaz-Ortiz, Á., Martín, D. R., & García, J. I. (2015). Influence of polarity and activation energy in microwave–assisted organic synthesis (MAOS). *Chemistry Open, 4*(3), 308–317.
68. Berlan, J., Giboreau, P., Lefeuvre, S., & Marchand, C. (1991). Organic-synthesis in microwave field-1st example of specific activation under homogeneous conditions. *Tetrahedron Letters, 32*(21), 2363–2366.
69. Saitou, K. (2006). Microwave sintering of iron, cobalt, nickel, copper and stainless steel powders. *Scripta Material, 54*, 875–879.
70. Chen, P. K., Rosana, M. R., Dudley, G. B., & Stiegman, A. E. (2014). Parameters affecting the microwave-specific acceleration of a chemical reaction. *The Journal of Organic Chemistry, 79*(16), 7425–7436.
71. Gupta, M., & Leong, E. W. W. (2008). *Microwaves and metals*. John Wiley & Sons.
72. An, N., Zhao, Y., Mao, Z., Agrawal, D. K., & Wang, D. (2018). Microwave modification of surface hydroxyl density for g-C3N4 with enhanced photocatalytic activity. *Materials Research Express, 5*(3), 035502.
73. Nozariasbmarz, A., Dsouza, K., & Vashaee, D. (2018). Field induced decrystallization of silicon: Evidence of a microwave non-thermal effect. *Applied Physics Letters, 112*(9), 093103.
74. Makul, N., & Agrawal, D. K. (2013). Microstructure and mechanical properties of microwave-assisted heating of pozzolan-Portland cement paste at a very early stage. *Songklanakarin Journal of Science & Technology, 35*(6), 693.
75. Upadhyaya, A., Tiwari, S. K., & Mishra, P. (2007). Microwave sintering of W–Ni–Fe alloy. *Scripta Materialia, 56*(1), 5–8.
76. Mondal, A., Upadhyaya, A., & Agrawal, D. (2010). Effect of heating mode on sintering of tungsten. *International Journal of Refractory Metals and Hard Materials, 28*(5), 597–600.
77. Prabhu, G., Chakraborty, A., & Sarma, B. (2009). Microwave sintering of tungsten. *International Journal of Refractory Metals and Hard Materials, 27*(3), 545–548.
78. Jain, M., Skandan, G., Martin, K., Cho, K., Klotz, B., Dowding, R., & Cheng, J. (2006). Microwave sintering: a new approach to fine-grain tungsten-I. *International Journal of Powder Metallurgy (Princeton, New Jersey), 42*(2), 45–50.
79. Wu, X., Thomas, J. R., Jr., & Davis, W. A. (2002). Control of thermal runaway in microwave resonant cavities. *Journal of Applied Physics, 92*(6), 3374–3380.
80. Kulumbaev, E. B., Semenov, V. E., & Rybakov, K. I. (2007). Stability of microwave heating of ceramic materials in a cylindrical cavity. *Journal of Physics D: Applied Physics, 40*(21), 6809.
81. Liao, S. Y. (1990). *Microwave devices and circuits*. Pearson Education India.
82. Tadmor, A. D., & Schachter, L. (1999). Optimized single-mode cavity for ceramics sintering. *IEEE Transactions on Microwave Theory and Techniques, 47*(9), 1634–1639.
83. Dube, C. L., Kashyap, S. C., & Dube, D. C. (2015). Growth mechanism of SiGe alloy nanowires synthesized in H-mode cylindrical cavity resonator. *Materials Discovery, 2*, 44–49.
84. Palaith, D., Silberglitt, R., Wu, C. C., Kleiner, R., & Libeld, E. L. (1988). Microwave joining of ceramics. *MRS Online Proceedings Library Archive, 124*, 255.
85. Cheng, J., Agrawal, D., Zhang, Y., Roy, R., & Santra, A. K. (2010). Synthesis of amorphous Si–Ge alloys using microwave energy. *Journal of Alloys and Compounds, 491*(1–2), 517–521.
86. Dube, D. C., Fu, M., Agrawal, D., Roy, R., & Santra, A. (2008). Rapid alloying of silicon with germanium in microwave field using single mode cavity. *Materials Research Innovations, 12*(3), 119–122.

87. Kashyap, S. C., Dube, C. L., Dube, D. C., & Agarwal, D. K. (2008). Synthesis of Si-Ge Alloys in MICROWAVE H-field. *Synthesis and Reactivity in Inorganic, Metal-Organic and Nano-Metal Chemistry, 38*(3), 264–266.
88. Cheng, J., Roy, R., & Agrawal, D. (2002). Radically different effects on materials by separated microwave electric and magnetic fields. *Materials Research Innovations, 5*(3–4), 170–177.
89. Cheng, J., Roy, R., & Agrawal, D. (2001). Experimental proof of major role of magnetic field losses in microwave heating of metal and metallic composites. *Journal of Materials Science Letters, 20*(17), 1561–1563.

Chapter 2
Microwave Processing Equipment and Modes

Abstract This chapter mainly covers microwave processing equipment and its components, such as microwave source, waveguides, circulators, and temperature measuring unit. The utility of each component is summarized along with the schematic diagram. This chapter describes how the equipment can be designed and used in multimode and single-mode operations for specific materials and applications. This chapter delineates types of applicators at length. The distribution of the E-and H-fields of electromagnetic radiation in applicators is discussed.

2.1 Introduction

Microwave processing is invariably carried out in a metallic chamber known as an applicator. The microwaves, generated by a magnetron source, are launched in the waveguide through an appropriate method: probe coupling, loop coupling, and aperture coupling [1]. Because of the presence of several modes in an applicator, the processing in this case is known as a multimode technique. There are, however, possibilities of exclusively using either E-or H-field associated with microwaves for the processing of certain materials. The deliberate selection of E-or H-field in an applicator leads to a single-mode technique. This chapter deals in detail with the microwave processing equipment required for carrying out MW processing in these modes.

2.2 Microwave Processing in Multimode Applicator

The multimode microwave processing is carried out in a multimode applicator; several modes are present inside the applicator depending upon its size. The most common example of multiple mode MW applicator is a domestic oven, having limited power of 100–1000 W. The domestic ovens are associated with a non-homogeneous distribution of the fields due to multiple reflections from the walls, formation of hot spots. The formation of hot spots is discussed in Sect. 1.4.1 of Chap. 1. Some

C. L. Dube, *Microwave Assisted Synthesis of Materials*, SpringerBriefs in Materials, https://doi.org/10.1007/978-3-032-01439-9_2

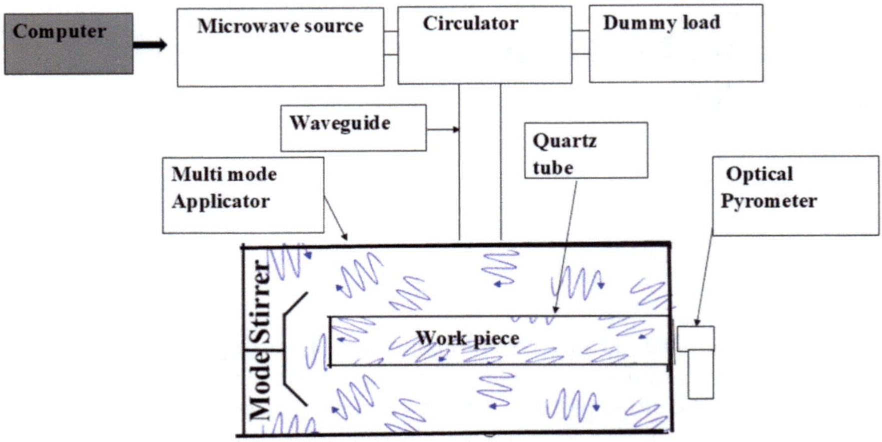

Fig. 2.1 A schematic diagram of a computer-interfaced MW processing unit with a multimode applicator

modifications in a domestic microwave oven for the synthesis of materials are suggested by various workers [2, 3]. The microwaves inside the applicator get reflected from the walls of the applicator and several modes are induced inside the applicator. The microwave energy is distributed among different modes. The energy stored in multiple modes interact with the workpiece as shown in Fig. 2.1. It mainly consists of MW source and a mode stirrer. Section 2.3 provides details on the types of MW sources, along with a brief overview of circulator, waveguide, dummy load, and optical pyrometer.

For uniform distribution of energy, the mode stirrer is attached in the multimode applicator, as shown in Fig. 2.1. The mode stirrer mixes the multiple modes of the electromagnetic waves, and to some extent, non-homogeneous distribution of MW field can be minimized. Multimode MW applicators are useful for high-throughput synthesis.

2.3 Microwave Processing in Single-Mode Applicator

A schematic diagram of a computer-interfaced MW processing unit with a single-mode applicator is shown in Fig. 2.2. For a single-mode microwave applicator, the size of the applicator is chosen in such a way that only a selected mode of electromagnetic radiation is present inside the applicator. In the single-mode applicator, predominately, E-or H-field is present at the workpiece and gives rise to either E-or H-field heating. The single-mode applicator can either be rectangular or cylindrical in shape. A schematic diagram of a computer-interfaced MW processing unit with a single-mode applicator is shown in Fig. 2.2.

It consists of a MW source, a waveguide, a circulator, an applicator, a dummy load (water), an optical pyrometer, and a set of tuners. Suitably adjusted shorting

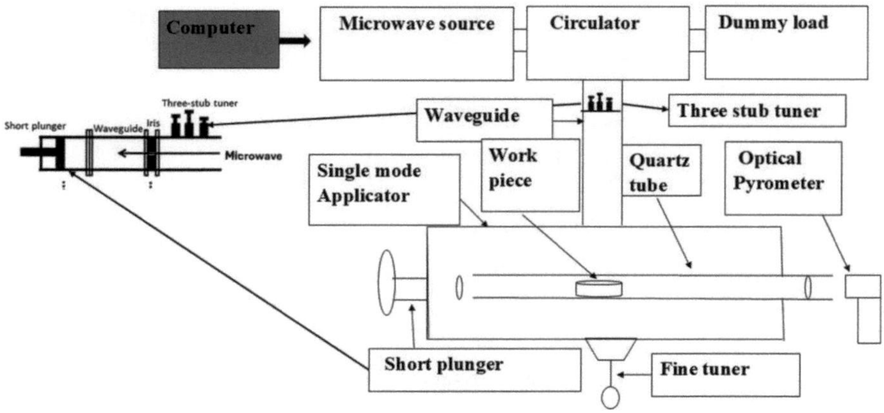

Fig. 2.2 A schematic diagram of a computer-interfaced MW processing unit with a single-mode applicator

plunger and an associated hardware (fine tuner) maximize the MW field in the centre of the workpiece and optimizes its coupling efficiency. If the load impedance varies during heating of the workpiece, it is necessary to keep the cavity tuned by frequent adjustment of the shorting plunger and fine tuner. A brief description of various parts/components of a processing unit (multi/single mode) is given below.

2.3.1 *Microwave Source*

To generate microwaves of desired high power and frequencies, microwave source is required. There are different kinds of sources used, such as magnetrons, klystrons, gyrotrons, and travelling wave tubes [1, 4]. Magnetrons are by far the most widely used microwave source, owing to their easy availability and low cost. The most commonly used magnetrons generate microwaves at a frequency of 2.45GHz and power typically in the range of 100–1000 W. The directional couplers are being used for measurement of MW power. The directional coupler in the unit is installed so that forward/reflected waves are separated and their power is measured independently. The purpose of a directional coupler is to measure microwave power, such as measuring incident power and reflected power.

2.3.2 *Circulators*

The materials which are not good absorbers of microwaves reflect a significant amount of the MW power. Consequently, a workpiece does not get the expected power and cannot be processed adequately. Besides, the reflected power can damage the generator. The use of circulator protects the microwave source. The

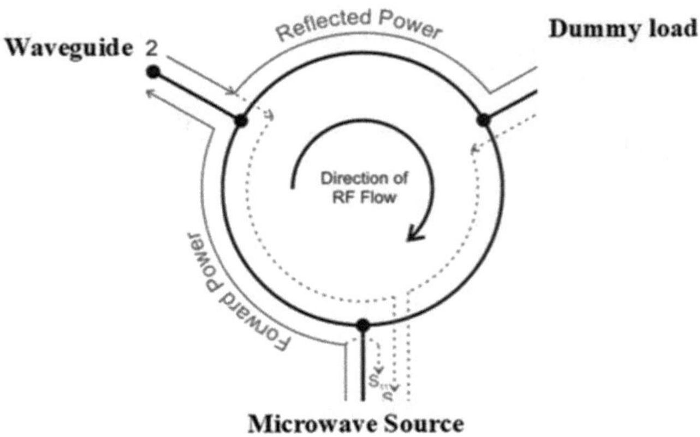

Fig. 2.3 The schematic diagram of a circulator

microwaves are allowed to pass through the circulator only in one direction. In a three-port circulator, one port is connected to the *microwave source*, another to the waveguide (connected to applicator), and the third port is connected to a dummy load (generally water). The reflected power towards the circulator is rerouted and absorbed by the dummy load (water). The schematic diagram of a circulator is shown in Fig. 2.3.

2.3.3 Waveguide

Waveguides are hollow tubes of specific shapes (with a circular or rectangular cross-section) and specific dimensions in which the electromagnetic waves propagate with a unique distribution of transverse and longitudinal distribution of the associated fields.

2.3.4 Modes in Waveguide

The mode in waveguide is a unique distribution of transverse and longitudinal components of the electric and magnetic fields of an electromagnetic radiation. There are different kind of modes, which can exist in a waveguide, namely.

Transverse Electromagnetic (TEM) Modes Both E-and H-fields are transverse to the direction of wave propagation. The component of E-and H-fields is zero, i.e. $E_z = H_z = 0$ [1, 5].

Transverse Electric (TE) Modes In the TE mode, the component of E-field along the x and y direction is transverse to the direction of propagation. The component of E-field is zero in the direction of propagation ($E_z = 0$) [1].

Transverse Magnetic (TM) Modes In the TE mode, the component of H-field along the x and y direction is transverse to the direction of propagation. The component of H-field is zero in the direction of propagation ($H_z = 0$) [1].

HE Modes In this case, neither the E-nor H-field transverses the direction of propagation. These are known as hybrid modes [1].

The most commonly used waveguides have rectangular or cylindrical cross-section. The modes used for materials processing are the TE and TM modes.

Rectangular and Cylindrical Waveguide The TEmnl/TMmnl modes are present in rectangular and cylindrical waveguides. The indices m, n, and l are the order of the modes, and they can be any integer, 0, 1, 2, etc. Depending upon the size of the waveguide, different modes exist in a waveguide. Each TE and TM wave in a waveguide has different field configurations. The different field configurations in TE and TM mode in rectangular waveguide are shown in Fig. 2.4.

For the TE mode, the electric field intensity is zero in the direction of propagation, and for the TM mode, the magnetic field intensity is zero. In a rectangular waveguide TE_{mnl}, m, n, and l represent the number of half-wavelengths (λ/2) the electric or magnetic field varies across the shorter (b-axis), longer (a-axis) dimensions of the rectangular waveguide and along the z-axis (the direction of propagation), respectively. In a circular waveguide TE_{mnl}, m, n represent the number of half-wavelengths along the circumference (m) and the number of half-wavelengths along the diameter (n) of the waveguide and l represents the number of half-wavelength variations along the z-axis (the direction of propagation). Almost entire energy is therefore confined in a cavity either in high E-or H-field resulting in more efficient interaction with certain materials. Figure 2.4 shows the field distributions of TE_{01} mode; electric and magnetic fields are

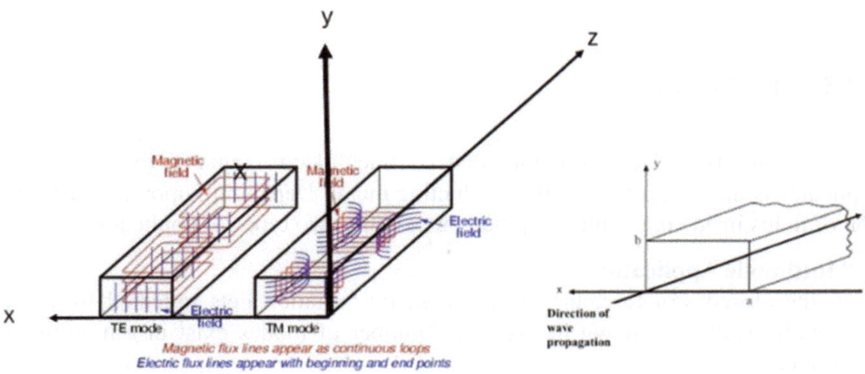

Fig. 2.4 The different field configurations in TE and TM mode in a rectangular waveguide

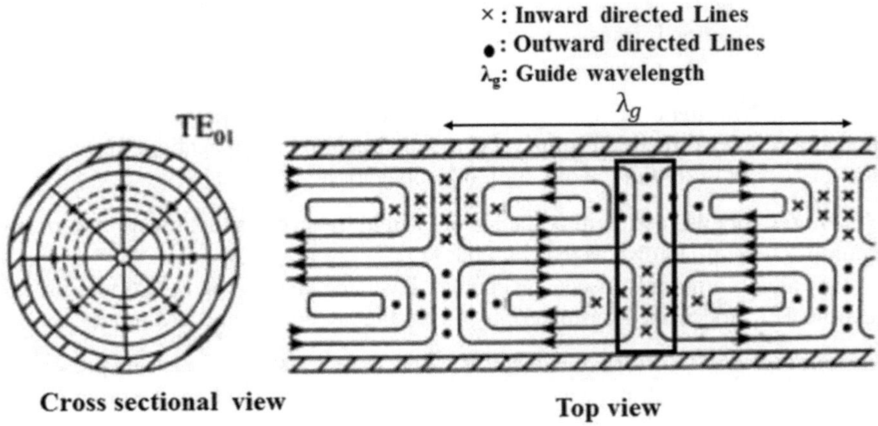

Fig. 2.5 Field distributions of TE_{01} mode in a cylindrical waveguide

represented by the solid lines and dashed lines, respectively. The guide wavelength (λ_g) is defined as the distance between two equal phase planes along the waveguide as shown in Fig. 2.2 [6]. It may be noted from Fig. 2.5 that predominantly H-field lines are present at the centre of the cavity (area enclosed in the small rectangular box).

The dominant mode in a particular waveguide is the mode having the lowest *cut-off frequency*. The *cut-off frequency* in a cylindrical waveguide is given by Eq. 2.1 [7].

$$f_{c_{\text{hollow}}} = \frac{c}{\pi D} J_{\text{mn}} \tag{2.1}$$

where J_{mn} is the *n*th root of the Bessel function of order m for the TM mode or the *n*th root of the differential coefficient of J_m for the TE mode, and D is a diameter cylindrical hollow waveguide.

2.3.5 Applicator

An applicator is a closed structure that confines electromagnetic radiation. The applicator is made up of perfectly conducting metal. Depending upon the distribution of fields inside the applicator, there are mainly two types of applicators.

1. **Multimode Applicator**
   In the closed structure, the electromagnetic radiation gets reflected from the metallic walls, as shown in Fig. 2.1. Number of modes exist in a multimode applicator.
2. **Single-Mode Applicator**
   A single-mode applicator is resonant cavity. It encloses a standing wave at specific frequency. Depending on the size of the applicator, a particular mode exists

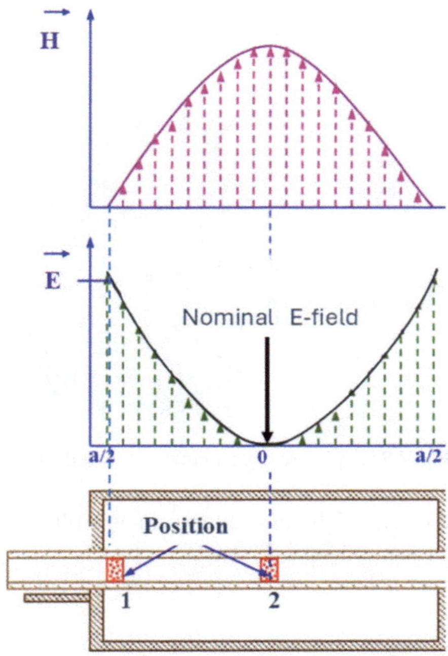

Fig. 2.6 The component of the electric field in the z direction is zero in TM mode in a cylindrical resonator of length *a*

in the applicator. It operates at frequencies above lowest cut-off frequency. The lowest cut-off frequency is explained below in Sect. 2.3.3. The strength of E-and H-field is very high in a single-mode applicator. Invariably, rectangular and cylindrical resonators are used for materials processing. The single-mode applicator is employed to study the role of E-and H-field in heating of the sample. If we assume wave is propagating in the Z direction as shown in Fig. 2. In the TM mode, a cylindrical resonator of length *a*, the component of the electric field in the direction of propagation of MW is zero, as shown in Fig. 2.6. It is worth to mention that at position 1 and position 2 (Fig. 2.6), the amplitude of H-field is minimum and maximum, respectively. Therefore, in the same applicator both the maximum and minimum H-field can be obtained for the materials processing. Besides, both the maximum and minimum E-field can also be obtained in the same applicator for the materials processing.

2.3.6 Short Plunger and Tuners

If the reflected power is high, the impedance matching of the MW processing unit needs to be adjusted/carried out appropriately/suitably with the help of a short plunger and different kinds of tuners such that maximum power is absorbed by the sample (load). Tuners such as *irises/three stub tuners/E-H plane tuners* are used for

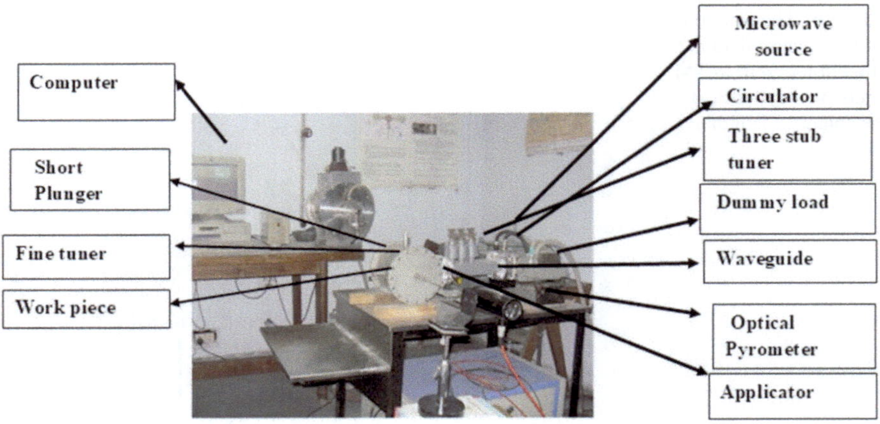

Fig. 2.7 A picture of computer-interfaced MW processing unit

the same. The placement of short plunger and tuners in a microwave processing is shown in Fig. 2.2.

2.3.7 Optical Pyrometer

To control and monitor the temperature of a workpiece, an optical pyrometer is used. The emitted radiation from a workpiece has an intensity distribution, the maximum of which can be detected by appropriately setting the emissivity (depending on the material, whose temperature is to be measured) of the pyrometer.

A typical picture of computer-interfaced MW processing unit is shown in Fig. 2.7.

2.4 Conclusion

A brief description of a MW processing equipment along with modes of operation is given in this chapter. In the multimode MW heating, several modes are present inside an applicator depending upon the applicator size. In the single-mode applicator, predominately, E-or H-field is present at the workpiece and gives rise to E/H field heating. The multimode MW unit is a general-purpose processing unit, whereas the single mode is employed for specific materials to obtain desired morphology.

References

1. Sadiku, M. N. (2015). *Principles of electromagnetics* (p. 2). Oxford university press Inc. First India edition.
2. Caddick, S. (1995). Microwave assisted organic reactions. *Tetrahedron, 51*(38), 10403–10432.
3. Villemin, D., & Thibault-Starzyk, F. (1991). Domestic microwave ovens in the laboratory. *Journal of Chemical Education, 68*(4), 346.
4. Harsany, S. C. (1996). *Principles of microwave technology*. Prentice-Hall, Inc..
5. Liao, S. Y. (1990). *Microwave devices and circuits*. Pearson Education India.
6. Ramo, S., Whinnery, J. R., & Whinnery, J. A. (1944). *Fields and waves in modern radio*. Wiley.
7. Webb, A. (2014). Cavity-and waveguide-resonators in electron paramagnetic resonance, nuclear magnetic resonance, and magnetic resonance imaging. *Progress in Nuclear Magnetic Resonance Spectroscopy, 83*, 1–20.

References

Chapter 3
Mechanism of Microwave–Material Interaction

Abstract This chapter gives detailed description of the microwave–material interaction. The classification of materials based on microwave–material interaction is presented. The different kinds of polarizations such as dipolar, interfacial, and electronic are elaborately incorporated in this chapter. The dielectric polarization losses, etc., responsible for volumetric heating are talked about in detail. The dependence of the relative permittivity and loss tangent of a material on frequency is covered. The concept skin depth and its role in MW heating of the metallic materials in powder form are discussed in detail.

3.1 Behaviour of Materials Under Microwaves

It is known that the whole range of materials such as metals, semiconductors, ferromagnetic materials, and insulators can be processed by employing MW energy. Microwave processing is attractive because it has many unique features, such as radically enhanced reaction kinetics, substantial enhancements in the sinterability, rapid joining of two materials (ceramic to ceramic, ceramic to metal, metal to metal), high reaction yields, structural uniformity of products, and accelerated nucleation due to both interaction of microwaves with the starting materials at atomic level and the presence of MW fields [1–5].

The behaviour of different materials when subjected to microwave exposure is dependent on their intrinsic properties, such as permittivity, permeability, and electrical conductivity [6]. Based on the interaction of the microwaves with a material, it is classified into the three categories, namely microwave reflectors, microwave transmitters, and microwave absorbers. The schematic diagram of MW interaction in these materials is shown in Fig. 3.1. It may be noted that the microwaves are either reflected, transmitted, or attenuated in a material.

C. L. Dube, *Microwave Assisted Synthesis of Materials*, SpringerBriefs in Materials, https://doi.org/10.1007/978-3-032-01439-9_3

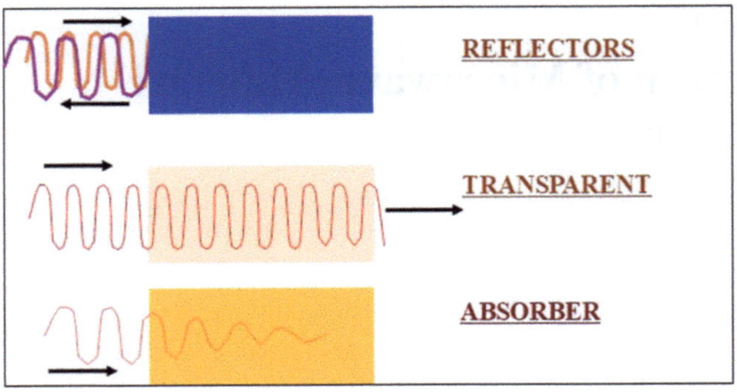

Fig. 3.1 The schematic of MW interaction in different kinds of materials

3.1.1 Microwave Reflectors

Bulk polycrystalline metals and alloys fall into the category of MW reflectors. The fields of the microwaves attenuate rapidly in the interior of a metal, as if the entire power is reflected back, and hence these materials are termed as MW reflectors.

At microwave frequencies, electromagnetic waves cannot penetrate deep inside a conductor due to skin effect. In fact, the waves die out beyond the skin depth. The skin depth is the distance at which the microwave power is reduced to 1/e of its value at the surface [7]. The skin depth *(d)* is frequency dependent and varies as per Eq. 3.1

$$d = \left(\frac{2}{\omega \mu \sigma} \right)^{1/2} \tag{3.1}$$

where ω is the angular frequency of electromagnetic radiation, μ is the permeability, and σ is the conductivity of a material. The schematic diagram of electromagnetic (EM) radiation penetration inside a powder material is shown in Fig. 3.2. The circles represent the constituent powder particles.

The skin depth for metals and alloys is a few micrometres (e.g. 1.5 μm for gold and 2.6 μm for brass). Due to very small skin depth, the bulk of polycrystalline metals and alloys do not interact with MW energy. However, it is possible to employ microwaves for the processing of metals and alloys when these are first prepared in a favourable morphology as discussed in Sect. 3.2.2.

3.1.2 Microwave Transparent Materials

Some materials, such as fused quartz, several glasses, ceramics (not containing any transition element), and teflon, neither interact with nor absorb incident MW radiation, i.e. these materials are transparent to microwaves. These transparent and

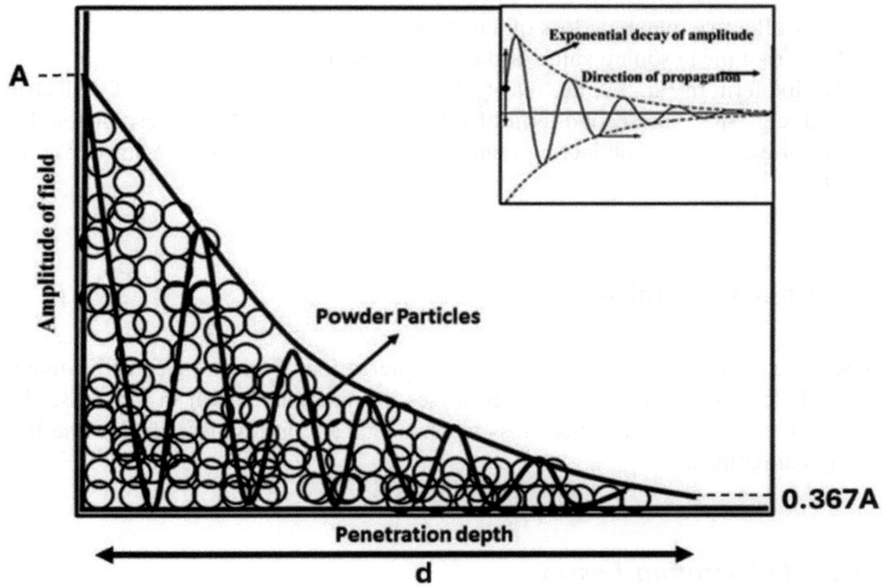

Fig. 3.2 A Schematic presentation of penetration of EM radiation inside a powder material

low-loss materials require special techniques for their interaction with microwaves. The MW transparent materials can be heated with the help of hybrid heating. The hybrid heating combines both MW and radiant heating as discussed in Chap. 1, Sect. 4.1.

3.1.3 Microwave Absorbers

These materials constitute the most important class of materials in MW processing which absorb energy from the microwave fields and get heated up rapidly. The examples include dielectrics, semiconductors, and magnetic oxides. During the material–microwave interaction in a MW absorbing material, the rate of energy absorption is expressed in terms of power per unit volume (P_a) as.

$$\mathbf{Pa} = \omega\varepsilon_o\varepsilon''_{eff}\boldsymbol{E}^2_{rms} + \omega\mu_0\mu''_{eff}\boldsymbol{H}^2_{rms} \tag{3.2}$$

The terms on the right side of the equation refer to electric and magnetic losses, respectively. Here, the symbols stand for the following:

$\boldsymbol{\omega}$ = angular frequency of the electromagnetic radiation (microwaves).

$\boldsymbol{\varepsilon_0}$ = permittivity of free space.

ε''_{eff} = effective relative dielectric loss of a material.

$\boldsymbol{E_{rms}}$ = root mean square value of the internal electric field.

$\boldsymbol{\mu_0}$ = magnetic permeability of free space.

$\mu_{eff}^{''}$ = effective magnetic loss of a material.

H_{rms} = root mean square value of the net internal magnetic field.

The first term of Eq. 3.2, i.e. $\omega\varepsilon_o\varepsilon_{eff}^{''}E_{rms}^2$ is responsible for heating of dielectric material, e.g. SiC. The second term of Eq. 3.1, i.e. $\omega\mu_0\mu_{eff}^{''}H_{rms}^2$ is responsible for heating of electrically conducting material, such as iron and copper powder compact [8].

3.2 Loss Mechanisms

Depending upon the nature of materials, the interaction of microwaves with a material leads to different kinds of energy losses in materials, which in turn raise the temperature of a material. These loss mechanisms are briefly described in the following paragraphs:

3.2.1 *Polarization Losses*

In the dielectric materials, the losses are due to dielectric polarization which involves short-range displacement of charges through formation and rotation of electric dipoles. The dielectric material under microwaves gets heated up due to dielectric heating of the sample. The dielectric heating depends on the material's complex permittivity (ε*). The complex permittivity can be expressed as.

$$\varepsilon^* = \varepsilon_0\left(\varepsilon_r^{'} - i\varepsilon_{eff}^{''}\right) \tag{3.3}$$

where ε_0 is the permittivity of free space, $\varepsilon_r^{'}$ is the relative dielectric constant, and $\varepsilon_{eff}^{''}$ is the effective relative dielectric loss factor. The real part of complex permittivity (ε _ r′) is a measure of the ability of a material to get polarized. The effective relative dielectric loss factor ($\varepsilon_{eff}^{''}$) is a measure of the ability of a material to convert the stored electromagnetic energy into heat. The loss tangent (tan δ) is used to express various losses in a dielectric material (Eq. 3.4) [8].

$$\tan\delta = \frac{\varepsilon_{eff}^{''}}{\varepsilon_r^{'}} \tag{3.4}$$

where $\varepsilon''_{eff} = \varepsilon''_s + \varepsilon''_d + \varepsilon''_i + \varepsilon''_e + \varepsilon''_c$, where ε''_s = space charge loss or interfacial polarization loss, ε''_d = dipolar polarization loss, ε''_i = ionic polarization loss, ε''_e = electronic polarization loss, and ε''_c = dc conductivity loss.

The dielectric constant is a measure of the net polarization of a material. There can be different kinds of polarizabilities in a material, such as dipolar, interfacial, electronic, and ionic [9], and all of these contribute to the effective relative dielectric loss. In addition to the frequency of microwaves, the polarizability depends on the temperature and composition of a dielectric material. The loss tangent increases with an increase in the temperature due to easy polarization. The loss tangent of a material is, therefore, a function of the frequency of the microwaves, the temperature/composition of the material, etc. A high loss tangent material is easily polarizable and hence is rapidly heated up in a microwave field as compared to a material having low-loss tangent. Different polarization/loss mechanisms are described in the following paragraphs.

3.2.1.1 Dipolar Polarization

A solid dielectric material can be polarized under an applied electric field as shown in Fig. 3.3. The polar dielectric materials have permanent dipole moments. The molecules having permanent dipole moment are randomly aligned in the absence of field due to thermal agitation and the net dipole moment of the material is zero. In the presence of an electric field, the dipoles tend to align themselves in the direction of the electric field, and the material is said to be polarized. Since the existing dipoles orient or say polarize in the direction of the applied field, this phenomenon is known as dipolar polarization. In an alternating electric field of the microwaves, the dipoles try to align in the direction of the field. However, dipoles do not have enough time to respond to the rapidly changing applied field and lag the applied field. That means a phase difference exists between the directions of the applied field and the dipoles. Due to phase difference, the energy is dissipated from the molecules upon random collision with the other agitating molecules and it results in dipolar heating of the entire material (volumetric heating). Another type of molecular interaction, e.g. rotation in liquid dielectric materials such as distilled water opposes the efforts made by dipoles to be in phase with E-field and leads to volumetric heating of the liquid dielectric materials. It may be noted/pointed out that the

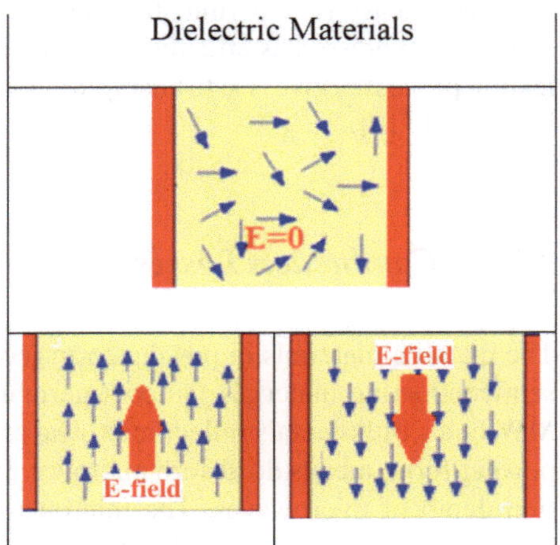

Fig. 3.3 The polarization of a dielectric material in an alternating E-field of microwave

nonpolar materials (homonuclear diatomic elements) are not a natural absorber of microwaves, thereby they cannot be heated effectively, e.g. melting wax will not work very well in microwaves.

3.2.1.2 Interfacial Polarization

In the absence of an electric field, there is no separation of opposite charges in a nonpolar material. Under the application of E-field of MW, the accumulation of free carriers/net charges at the grain boundaries, defects, inhomogeneities, etc. results in the interfacial polarization of a material. The extent of polarization depends on the strength of the applied electromagnetic field.

3.2.1.3 Electronic Polarization

An atom is electrically neutral and does not have dipole moment in the absence of an externally applied electric field. On the application of the external field, the valance electrons shift with respect to the ionic cores in a direction opposing the field and result in an induced dipole moment. In a covalently bonded solid, the electronic polarization results in a large dielectric constant, e.g. silicon exhibits a dielectric constant approximately 11.7 at 2.45 GHz.

3.2.1.4 Ionic Polarization

In the neutral states, the net dipole moment of an ionic molecule/solid is zero. Under the application of a MW field, the positive and negative ions, within the molecule, are relatively displaced and net dipole moment is no longer zero.

The induction of polarization in a material can contribute to the losses, and it is not always easy to experimentally differentiate between the loss mechanisms; the losses are expressed as effective relative dielectric loss, ε''_{eff} in Eq. 3.2. These polarization processes give rise to absorption losses in a certain frequency range as shown in Fig. 3.4 [10].

3.2.2 Conduction Losses

The metallic materials in powder form (copper, iron, cobalt, and tungsten etc.) and semiconducting materials (silicon, germanium etc.) get efficiently heated under a MW field. Bulk metals reflect microwaves following Maxwell's law; however, in a powder form, metals do absorb microwaves, if particle size is less than skin/penetration depth of microwaves. The microwave field attenuates sharply and results in spatially varying electric field in a conducting material, which in turn induces a

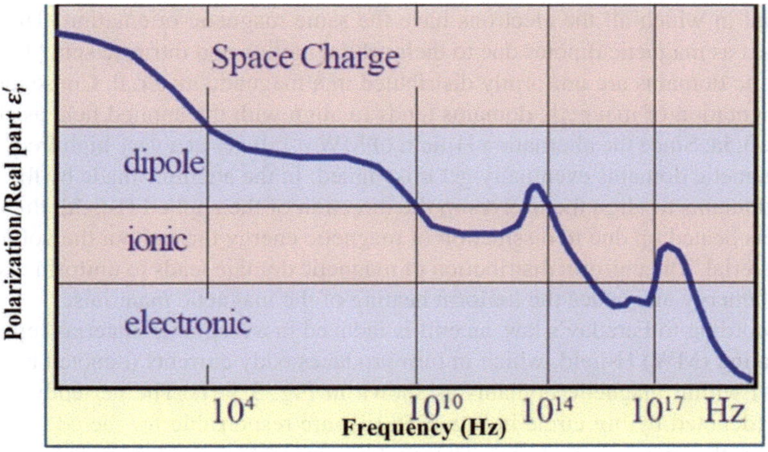

Fig. 3.4 A typical frequency dependence of the relative permittivity

time-varying magnetic field (in a direction opposite to that of applied MW H-field) as per Faraday's law [11]. A time-varying magnetic field generates an electric field, and this electric field then acts on the electron, accelerating it and increasing its kinetic energy. Thereby, the kinetic energy gained by free electrons under the time-varying H-field gets converted into heat and thus heating of entire metallic powder materials takes place due to conduction losses.

The free electrons in metals are easily affected by an applied electric field and accelerated. However, the scattering with the ions impedes these electrons and results in the ohmic losses and the production of heat, called Joule heating (conduction losses). In the case of semiconductors, e.g. Si and Ge, as the temperature rises, more free carriers are generated and the conduction losses become significant. This phenomenon essentially accounts for the higher MW absorption and faster heating behaviour of Si and Ge [12]. In addition, the eddy currents and magnetic induction also play an important role in heating certain resistive/semiconducting materials.

3.2.3 *Magnetic Losses*

It is known that the charged particles (electrons and ions.) do not absorb energy from a magnetic field. In some materials, however, the magnetic dipoles may couple with the magnetic component of the electromagnetic field and provide an additional heating mechanism. The MW heating of magnetic materials takes place mainly via hysteresis loss, eddy current loss, and ferromagnetic resonance loss.

The hysteresis loss in a magnetic material occurs due to the interaction between alternating H-field of the applied MW and the net magnetic field of a magnetic domain in the material. A magnetic domain is defined as a region within a magnetic material in which all the electrons have the same magnetic orientation. The electrons act as magnetic dipoles due to their orbital motion and intrinsic spin [13]. The magnetic domains are uniformly distributed in a magnetic material. Consequently, the orientation of magnetic domains tends to align with the applied field as shown in Fig. 3.5a. Since the alternating H-field of MW oscillates at a very high frequency, the magnetic domains eventually get misaligned. In the attempts made by the magnetic domains to align themselves in the direction of the applied H-field, the material gets heated up due to dissipation of magnetic energy throughout the volume of the material. The uniform distribution of magnetic domain leads to uniform dissipation of energy and hence the uniform heating of the magnetic materials.

According to Faraday's law, an emf is induced in a magnetic material kept in an alternating (MW) H-field, which in turn produces eddy currents (denoted by small circles) within magnetic domains as shown in Fig. 3.5b(i). The net eddy current loops (denoted by big circle in Fig. 3.5b (ii)) are responsible for the eddy current losses in the magnetic material. These eddy current losses, like hysteresis losses, result in heating of the entire magnetic material.

There are other loss mechanisms such as domain wall resonance and ferromagnetic resonance losses operative in the magnetic materials and leading to the heating of these materials. However, it is not always easy to differentiate experimentally between the loss mechanism operative simultaneously in a magnetic material. The losses are, therefore, typically expressed as effective magnetic losses.

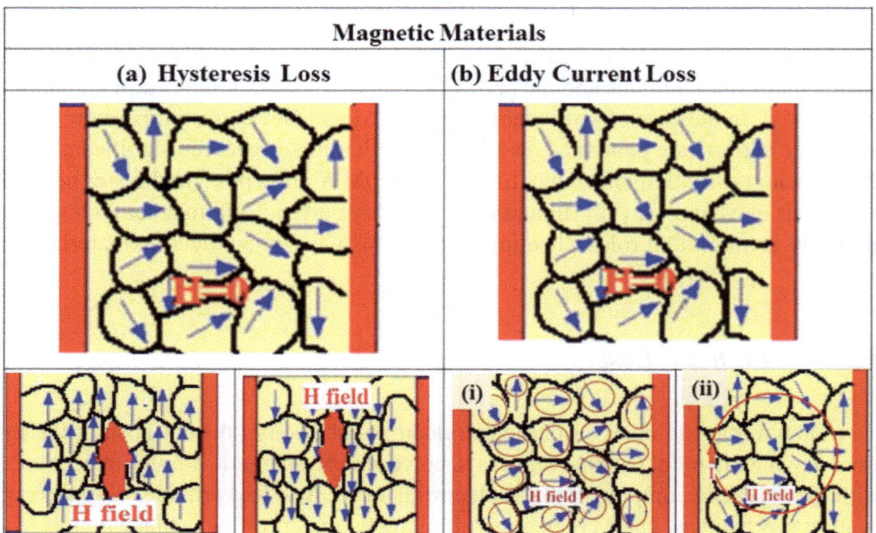

Fig. 3.5 (**a**) The alignment of magnetic domains. (**b**) Induction of an eddy current loop in a magnetic material in an alternating (MW) H-field

3.3 Conclusions

This chapter mainly emphasized the behaviour of materials under microwaves. The materials are classified as reflectors, transparent, and absorbers to the microwaves. The different kinds of materials interact in a very distinctive manner and thereby leading to different kinds of losses, such as dielectric polarization losses and magnetic losses. In the process of polarization, the frictional heat gets generated, which is responsible for efficient heating of the material. In a magnetic material, however, the hysteresis, eddy current, and ferromagnetic resonance losses are found responsible for the heating of the material.

References

1. Baghurst, D. R., Chippindale, A. M., & Mingos, D. M. P. (1988). Microwave syntheses for superconducting ceramics. *Nature, 332*(6162), 311–311.
2. Ramesh, P. D., & Rao, K. J. (1995). Microwave-assisted synthesis of aluminum nitride. *Advanced Materials, 7*(2), 177–179.
3. Landry, C. C., & Barron, A. R. (1993). Synthesis of polycrystalline chalcopyrite semiconductors by microwave irradiation. *Science, 260*(5114), 1653–1655.
4. Freeman, S. A., Booske, J. H., & Cooper, R. F. (1998). Modeling and numerical simulations of microwave-induced ionic transport. *Journal of Applied Physics, 83*(11), 5761–5772.
5. Dube, D. C., Agrawal, D., Agrawal, S., & Roy, R. (2007). High temperature dielectric study of Cr_2O_3 in microwave region. *Applied Physics Letters, 90*(12), 124105.
6. Mishra, R. R., & Sharma, A. K. (2016). Microwave–material interaction phenomena: heating mechanisms, challenges and opportunities in material processing. *Composites Part A: Applied Science and Manufacturing, 81*, 78–97.
7. Mondal, A., Agrawal, D., & Upadhyaya, A. (2010). Microwave sintering of refractory metals/alloys: W, Mo, Re, W-Cu, W-Ni-Cu and W-Ni-Fe alloys. *Journal of Microwave Power and Electromagnetic Energy, 44*(1), 28–44.
8. Rybakov, K. I., Semenov, V. E., Egorov, S. V., Eremeev, A. G., Plotnikov, I. V., & Bykov, Y. V. (2006). Microwave heating of conductive powder materials. *Journal of Applied Physics, 99*(2), 023506.
9. Griffiths, D. J. (2015). *Introduction to electrodynamics*. Pearson India Education Services.
10. Kingery, W. D., Bowen, H. K., & Uhlmann, D. R. (1976). *Introduction to ceramics*. John Wiley & Sons.
11. Dube, C. L., Kashyap, S. C., Dube, D. C., & Agrawal, D. K. (2013). Rapid alloying of tetrahedral diamagnetic semiconductors in microwave H field. *Journal of Alloys and Compounds, 571*, 75–78.
12. Vaucher, S., Catala-Civera, J. M., Sarua, A., Pomeroy, J., & Kuball, M. (2006). Phase selectivity of microwave heating evidenced by Raman spectroscopy. *Journal of Applied Physics, 99*(11), 113505.
13. Cullity, B. D., & Graham, C. D. (2011). *Introduction to magnetic materials*. John Wiley & Sons.

Chapter 4
Microwave Processing of Semiconducting Materials

Abstract This chapter deals with the microwave processing of a powder mixture of silicon and germanium (group IV semiconductors) which has resulted in the growth of nanostructures of $Si_{1-x}Ge_x$ ($0 \leq x \leq 0.25$) alloys. The role of powder particle size and its shape are found to be crucial in the formation of these nanostructures by employing the MW processing in single mode. The reaction atmosphere—air or Argon—has played a definitive role in controlling the morphology of the resulting semiconducting alloys (nanoparticles and nanowires). This technique of material processing has reduced the equilibrium temperature of SiGe alloy formation considerably. A new mechanism of interaction of MW with semiconductors has been briefly described at the end of this chapter.

4.1 Introduction

4.1.1 Group IV Semiconductors

Nanostructures of semiconducting materials have gained momentum due to their unique and improved features as compared to their bulk counterparts. The nanostructures carry charge and excitons efficiently and are therefore potentially ideal building blocks for nanoscale electronics and optoelectronic [1]. Silicon and germanium are tetrahedrally coordinated indirect band gap semiconductors with band gaps of 1.12 and 0.69 eV, respectively. They form a continuous series of random substitutional solid solutions over the entire composition range [2]. The band gap of $Si_{1-x}Ge_x$ alloys (where $0 \leq x \leq 1$) varies with change in germanium concentration in the alloy; it has a dominant effect on the optical and electrical characteristics of these alloys. Besides the use of single crystalline $Si_{1-x}Ge_x$ alloys in photodetectors and thermo-generators, its polycrystalline form finds application in microelectronic devices, transistors, and micro-electromechanical systems (MEMS) [3]. In addition, the nanowires of this alloy exhibit good thermoelectric properties [4].

To synthesize $Si_{1-x}Ge_x$ alloy of different compositions, desired amounts of Si and Ge powders were thoroughly mixed and pelletized. The MW processing

C. L. Dube, *Microwave Assisted Synthesis of Materials*, SpringerBriefs in Materials, https://doi.org/10.1007/978-3-032-01439-9_4

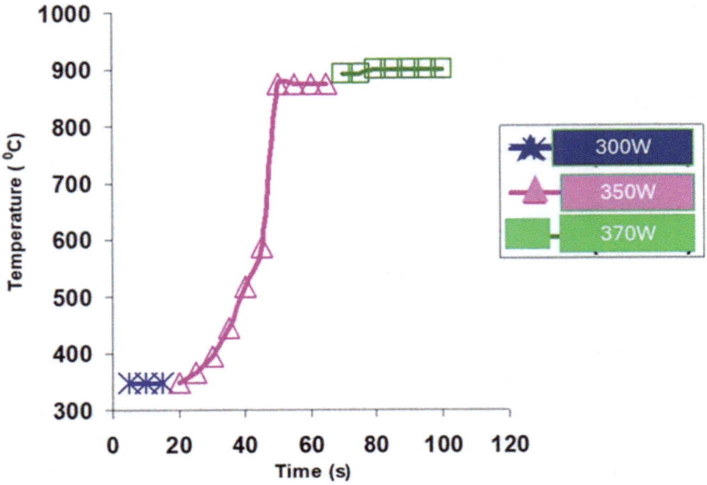

Fig. 4.1 A typical heating profile in a single-mode MW applicator

temperature of the starting powder mixture was kept for a set period (2–10 min). A pellet was located at the centre of a single-mode cylindrical applicator operated at 2.45 GHz and ~ 450 W. The dimension of the applicator was such that the TE_{011} mode can exist in the applicator. The H-field is maximum at the centre of the applicator (position 2 in Fig. 2.4 of Chap. 2). Once the microwave power was put on and the applicator got tuned, temperature of the pellet raised to a desired value. A typical heating profile is shown in Fig. 4.1.

The processing is carried out in an argon (Ar) environment to avoid the formation of undesired compounds during the process [5].

The temperature acquired by the sample was measured using an optical pyrometer (Raytek model RAYMA2SCSF) at an emissivity of 0.6 [6]. The ambience of materials' synthesis has a significant role in tailoring the morphology and hence physical/microstructural properties of the final product. This is demonstrated by processing Si and SiGe alloys in both argon and air ambience [7, 8].

4.2 Microwave Processing in Argon Ambience

The MW processing temperature of the precursors (mixture of silicon and germanium powder) was maintained for a stipulated period (2–5 min) [9]. The pellets of the starting mixture $Si_{0.95}Ge_{0.05}$ (named as 5SG) were MW processed in different runs at 800 °C for 2 min, 900 °C for 2 min, and 900 °C for 5 min. The MW-processed pellets were named as 5SG8002, 5SG9002, and 5SG9005, respectively. The resulting polycrystalline samples were analysed for their parameters/properties by using several techniques [10].

4.2.1 Structural Analysis

The structural/phase analysis of the starting mixture and of the MW-processed pellets was carried out by employing X-ray diffractometer in the powder mode. Figure 4.2 shows X-ray diffraction pattern (XRD) of these samples.

The XRD pattern of starting mixture (sample 5SG) shows the well-defined diffraction peaks corresponding to both the stating constituents (Si and Ge), establishing their elemental form to begin with. The lattice parameter *a* of Si and Ge has been obtained as 5.42 and 5.65 Å, respectively. The XRD peaks in the other diffractograms, recorded for the MW-processed pellets, belong to the diamond structure and reveal anticipated shift in (111), (220), (311), (400), and (331) reflections from Si lattice towards lower 2θ, which establishes the alloy formation along with changed lattice parameter (due to alloy formation). However, a few peaks in the XRD scan of the 5SG8002 marked with & (corresponding to unreacted Ge in the sample) are also recorded. The shape of the peak in the XRD scan of the 5SG8002, processed at 800°C (shown in the encircled area) does not fit Gaussian function, which in turn indicates that MW synthesized alloy may not be a single phase [9]. It may, however,

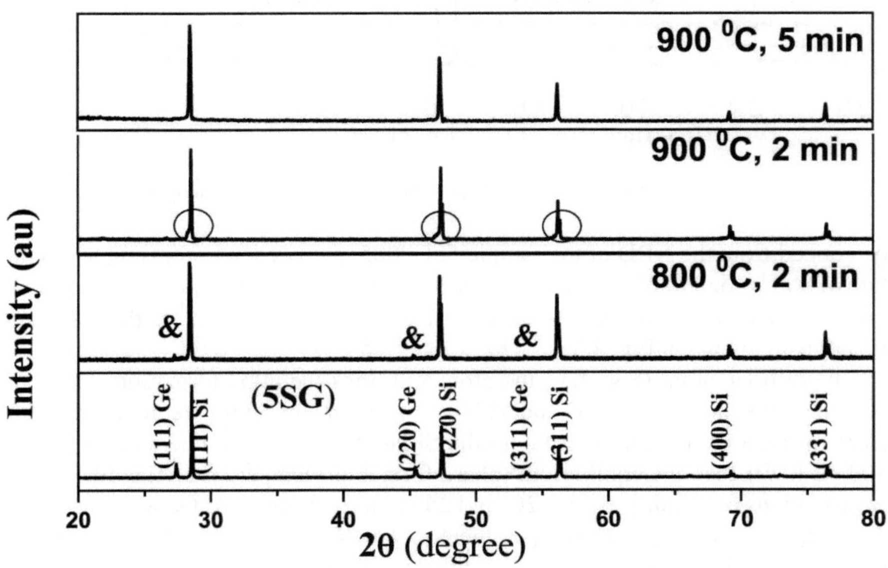

Fig. 4.2 The XRD pattern of 5SG and microwave-processed samples

Fig. 4.3 The XRD pattern of 5SG9005 sample replotted in between diffraction angle 27 and 30°

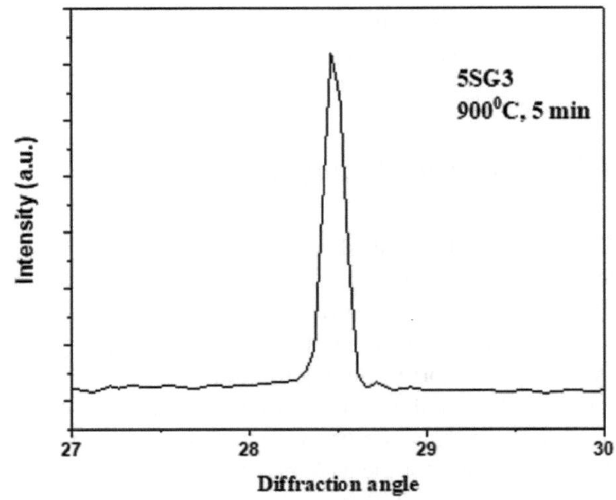

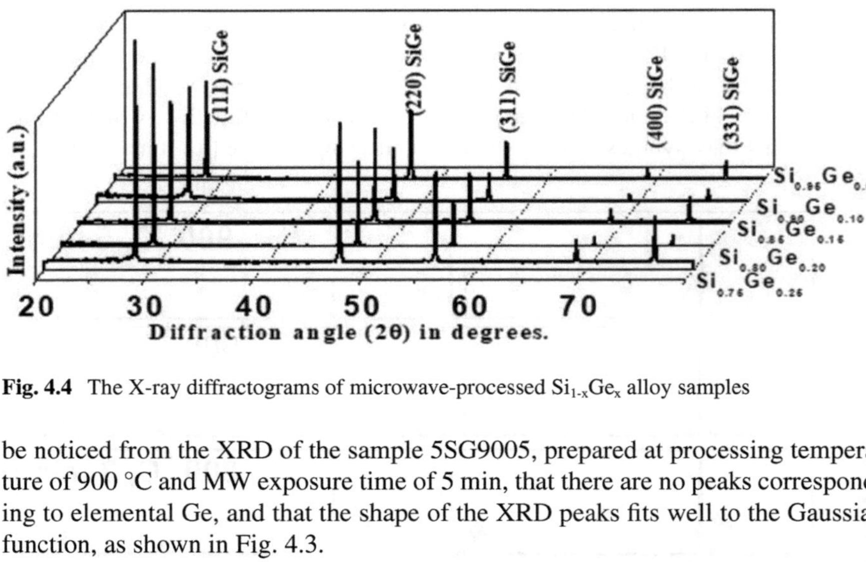

Fig. 4.4 The X-ray diffractograms of microwave-processed $Si_{1-x}Ge_x$ alloy samples

be noticed from the XRD of the sample 5SG9005, prepared at processing temperature of 900 °C and MW exposure time of 5 min, that there are no peaks corresponding to elemental Ge, and that the shape of the XRD peaks fits well to the Gaussian function, as shown in Fig. 4.3.

The diffractogram (Fig. 4.3), therefore, confirms the alloy formation. The processing temperature of 900 °C and MW exposure time of 5 min are therefore sufficient for total incorporation of Ge in the Si lattice [9]. This temperature is nearly 400 °C lower than the equilibrium (phase formation) temperature. The other alloy compositions containing 10, 15, 20, and 25 atomic weight % of Ge were also synthesized at the same processing parameters (900 °C, 5 min). The X-ray diffractograms (Fig. 4.4) confirm the alloy formation in all these cases.

4.2.2 Microstructural Analysis

To investigate the effect of MW processing parameters and composition on the microstructure of synthesized alloy samples, electron microscopic (SEM) study was performed.

4.2.2.1 Effect of Alloy Composition on its Microstructure

The micrographs of $Si_{1-x}Ge_x$ ($0 \leq x \leq 0.25$) alloys of various compositions synthesized at 900 °C in 5 min are shown in Fig. 4.5.

It can be seen from the scanning electron microscope (SEM) pictures that as the germanium content is increasing in the alloy composition, the samples show signature of enhanced melting. The encircled areas in Fig. 4.5 show features of melting in the alloy samples. It is understandable from the equilibrium phase diagram of Si and Ge (Fig. 4.6) that the melting temperature of different alloys decreases with an increase in the germanium content in the sample.

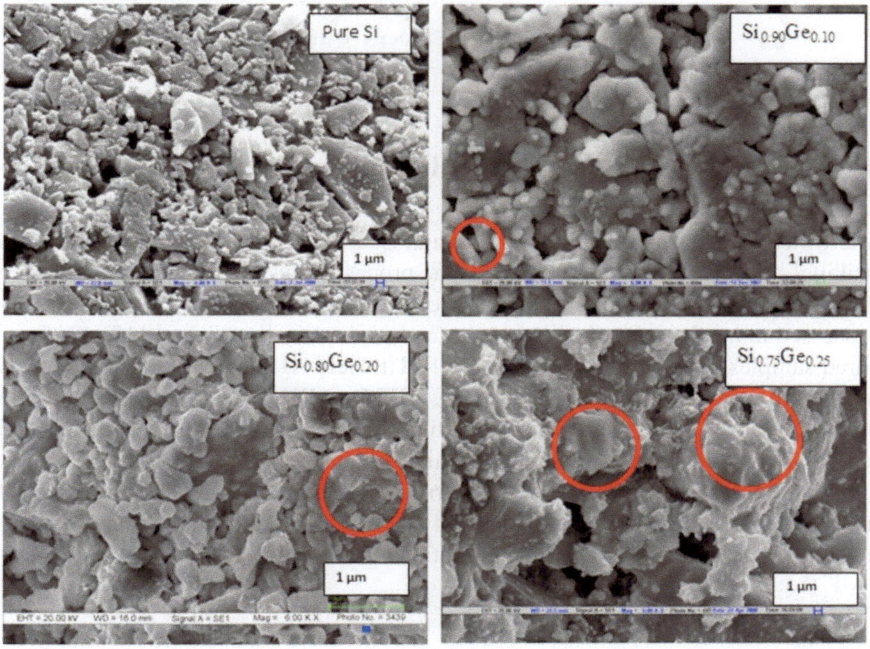

Fig. 4.5 Scanning electron micrographs of $Si_{1-x}Ge_x$ ($0 \leq x \leq 0.25$) alloys of different compositions (MW processed at 900 °C for 5 min)

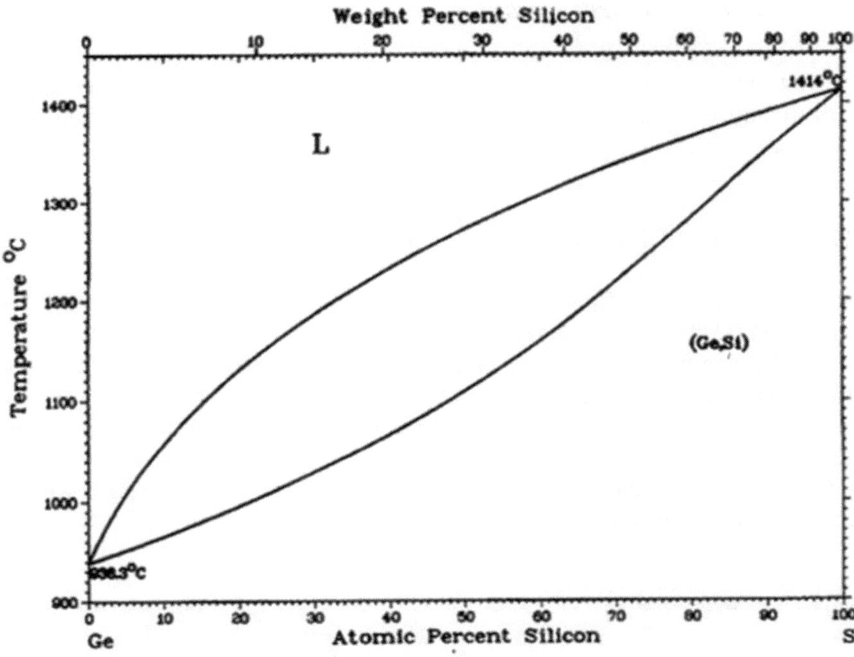

Fig. 4.6 Ge-Si phase diagram [11]

4.3 Microwave Processing in Air Ambience

Microwave processing of a mixture of Si and Ge powder in air ambience resulted in strikingly different morphology, unlike the processing in Ar ambience. To prepare the $Si_{1-x}Ge_x$ alloys of different compositions, mixtures of Si and Ge powder were processed in an air ambiance in a single-mode cylindrical applicator [10]. The prepared samples were labelled as AICCSGTTTtt. Here AI stands for air ambient CC for atomic weight % of Ge, SG for SiGe, TTT for MW processing temperature in three digits in °C, and tt for processing time in min. For example, AI05SG90005 implies that the sample is processed in air ambient with 5 at. wt.% of Ge and at temperature of 900 °C for 5 min. The structural and microstructural properties of the samples of different compositions and synthesized at different process parameters are described in the following subsections.

4.3.1 Structural Analysis

The XRD pattern of the nanowire confirms the diamond structure and hence alloy formation (Fig. 4.7). The lattice parameter *a* of the $Si_{0.75}Ge_{0.25}$ alloy nanowire is found to be 5.47 Å which understandably lies between lattice parameters *a* value of the two constituents.

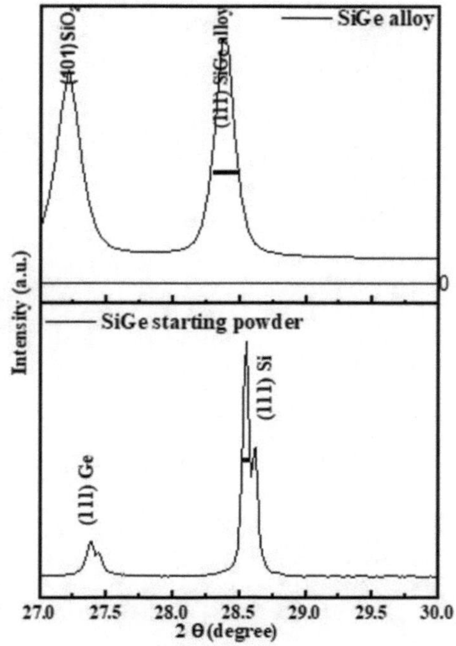

Fig. 4.7 The XRD patterns of starting powder and alloy nanowires

The full width at half maximum of the peak corresponding to (111) diffracting planes for AI25SG90010 is found to be 3.5 times that of the corresponding peak in the XRD of the starting powder. This increase in FWHM can be ascribed to the reduction in the size of a diffracting crystallite [12], i.e. to the nanowires (in the present case).

4.3.2 *Microstructural Analysis*

4.3.2.1 Effect of MW Processing Temperature and Duration on Microstructure

To study the effect of MW processing temperature and duration, the pellets containing 15 and 25 atomic weight % Ge were processed at different temperatures and for different durations. The samples containing 15 atomic weight % of Ge were processed at a temperature of 950 °C for 5, 10, or 15 min and their micrographs are shown in Figs. 4.8a–c. It is clearly seen from these micrographs that the length of the nanowires increases with MW processing time, which is expected. Similar findings are also concluded from the micrographs (Figs. 4.8e, f) of the samples AI25SG9002 and AI25SG90010.

These micrographs, in addition, reveal that as the MW processing temperature is increased from 800 to 900 °C, the nanowires grow thinner. Furthermore, Fig. 4.8d shows small nanowires originating from various alloy particles. The inset of

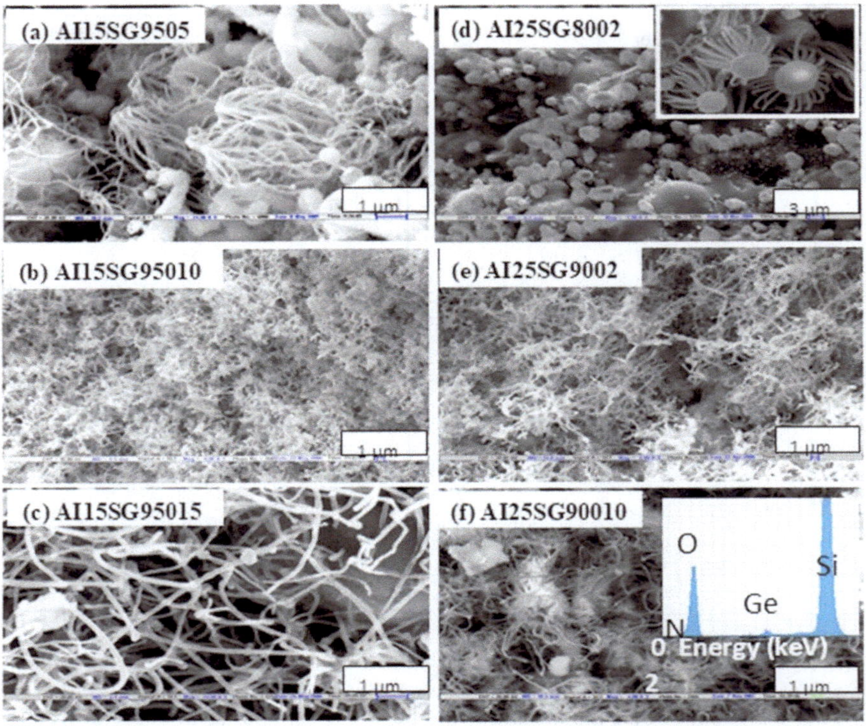

Fig. 4.8 SEM pictures of (**a**) AI15SG9505, (**b**) AI15SG95010, (**c**) AI15SG95015, (**d**) AI25SG8002, (**e**) AI25SG9002, and (**f**) AI25SG90010 samples

Fig. 4.8d shows a magnified view of a part of the micrograph. It may be noted from Figs. 4.8b, c–f that almost all the alloys are transformed into nanowires. The energy dispersive analysis of X-rays spectrum of the sample (AI25SG90010), shown in the inset of Fig. 4.8f, indicates that the nanowires mainly comprise Si and Ge with a small fraction of nitrogen and oxygen. This could be due to the formation of oxide/oxynitride on the exterior of nanowires grown in the air ambience. The X-ray photoelectron spectroscopy analysis confirmed the presence of Si and Ge in the sample along with formation of oxides/oxynitrides of Si and Ge on the surface of the nanowire sample [10].

It is suggested that the changing local magnetic field induces circulating nonconservative electric field which acts as a driving force for growth of nanowires.

The appearance of spherical features, seen in Fig. 4.8d, can be understood on the basis that initially in the process of surface energy minimization, a spherical surface encloses maximum volume for a given surface area.

Figure 4.9a displays a high-resolution transmission electron micrograph (HRTEM) of AI25SG90010 sample. The Z contrast (Fig. 4.9a) in the HRTEM image clearly implies uniform distribution of Ge in the MW-processed sample. The

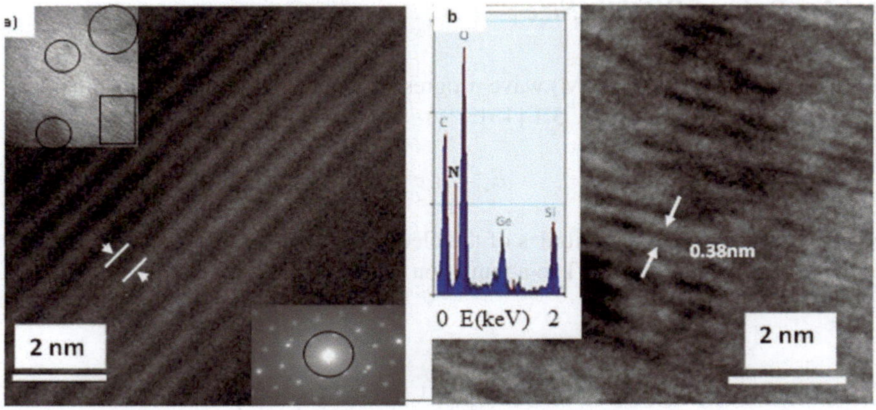

Fig. 4.9 HRTEM picture of (**a**) AI25SG90010 and (**b**) AI10SG9005 samples

planes in the core of a nanowire exhibit a separation of nearly 5.5 Å, which equates well with the interplanar spacing in (100) of $Si_{0.75}Ge_{0.25}$ alloy. Furthermore, the selected area diffraction pattern (inset of Fig. 4.9a) confirms the crystalline nature of the nanowires.

The high-resolution transmission electron micrograph (HRTEM) corresponding to sample AI10SG9005 is shown in Fig. 4.9b. The lattice spacing is found to be 0.38 nm, which matches well with the interplanar spacing in (110) of $Si_{0.90}Ge_{0.10}$ alloy.

The MW processing of the mixture of silicon and germanium in air showed the random growth of SiGe alloy nanowires with very high yield. Most of the SiGe alloy nanowires are extremely long (>10 μm) and randomly oriented. The Z contrast of grown nanowires, seen in the high-resolution transmission electron micrograph (HRTEM) confirmed the uniform distribution of germanium in the nanowire.

It was proposed that the oxygen, present in the air ambient during MW processing, assisted the growth of the nanowires. It was also found that the growth is promoted in the alloy samples having higher concentration of germanium. Possibly the higher vapour pressure of Ge, as compared to that of silicon, at a growth temperature was responsible for the enhanced growth of nanowires [10].

4.4 Role of Particle Size and Particle Shape in SiGe Alloy Formation

The particle size and shape play an important role in the absorption of microwaves by certain materials like metals and semiconductors. The effect of both these factors on the alloying of SiGe is discussed briefly in the following subsections:

4.4.1 Role of Particle Size

When an electromagnetic (EM) wave progresses in a lossy medium, it is attenuated exponentially as given by Eq. 3.1 [13].

$$E_x = E_o e^{-i\alpha x} \tag{4.1}$$

where E_x and $E_o$are the amplitudes of the electric field at depth x and at the surface of the medium, respectively. The attenuation factor, α, is expressed as

$$\alpha = \omega\left(\frac{\mu_0\mu'\varepsilon_0\varepsilon'}{2}\right)^{\frac{1}{2}}\left[\left(1+\left(\frac{\varepsilon_{eff}''}{\varepsilon'}\right)^2\right)^{\frac{1}{2}}-1\right]^{\frac{1}{2}} \tag{4.2}$$

where the symbols μ' and ε' stand for the permeability and permittivity of the medium, respectively, and $\varepsilon_{\mathrm{eff}}$''is the effective relative dielectric loss factor of the medium. As already discussed in Chap. 3 under Sect. 3.1.1 that the distance from the surface of the material at which magnitude of field strength reduces to 1/e of its initial value at the surface is known as skin depth (d) and is defined as d = 1/ α [14] (Chap. 3, Sect. 3.1.1).

In the case of diamagnetic semiconducting materials such as silicon and germanium, the skin depth is of the order of 0.1 mm in GHz range [15]. The diamagnetic materials such as Si and Ge have small susceptibility values (very small or equal to 0). Therefore, heating due to the magnetic losses is ruled out in the diamagnetic materials. However, it has been reported that size of a particle in nanoscale can play a crucial role in inducing magnetic moment in a diamagnetic material in bulk form [16]. Since the size of particles in the starting mixture of Si and Ge powder is less than 80 nm (Fig. 4.10) and is much smaller than the skin depth, it leads to the most effective interaction of the MW with these semiconductors. The inset of Fig. 4.10 shows an EDAX spectrum recorded for the starting mixture at different locations on the same surface of a pellet confirms uniform distribution of starting constituents. This spectrum also reveals the presence of oxide impurities which suggest that each individual semiconductive powder particle in the compact is surrounded by an oxide layer.

The presence of such oxide shells on individual semiconductive silicon and germanium particles significantly increases the electromagnetic power absorption in the powder compact as skin depth of microwave in oxide of silicon and germanium is larger compared to non-oxidized semiconductive powders. However, these oxide shells decompose at higher temperature [17].

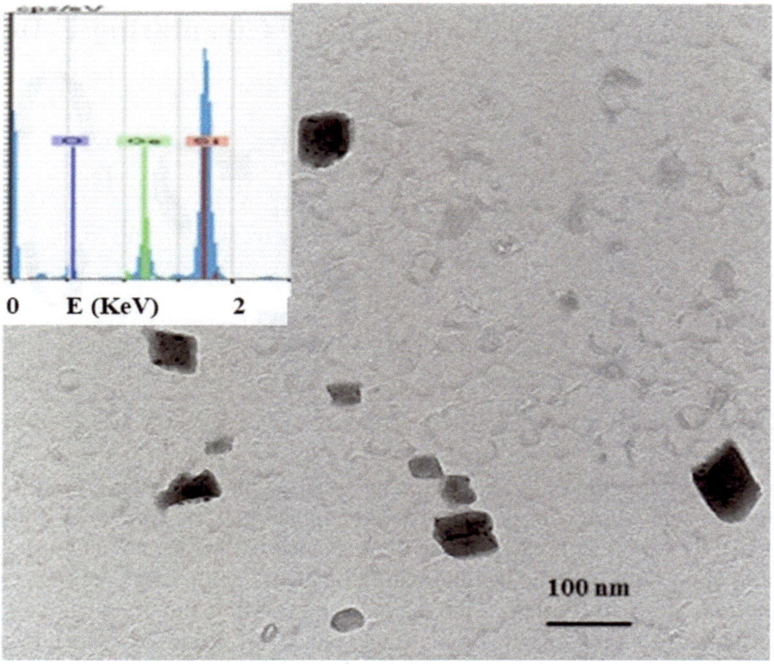

Fig. 4.10 TEM image and EDAX spectrum of constituents (Si and Ge). In a starting powder

4.4.2 Role of Particle Shape

It may be noted from the TEM image (Fig. 4.10) of the starting mixture that the constituent particles (Si and Ge) are of irregular shapes. During microwave processing of the starting materials, the random shaped particles of (Si + Ge) get exposed to uniform H-field lines in the single-mode applicator. MW interaction changes due to change in shape of the particles during alloy formation and the growth of the alloy particles; the oscillatory H-field of applied MW will induce an electric field and hence the induced surface currents flow across the boundary of the irregularly shaped particles [8]. The magnitude of the surface currents varies from particle to particle depending upon its size and shape. The *variation in current* owing to Lenz's law forces the current loops to oppose the change by absorbing energy from the MW, which in turn generates heat and raises the temperature of the sample.

If the MW field is on, the temperature of the pellet keeps on increasing and thereby surface energy of the particles increases. To minimize surface energy, the particles acquire spherical shape of different sizes and thus their magnetic permeability changes [18]. The magnetic induction in these particles, exposed to uniform H-field lines, varies from particle to particle and the local circulating nonconservative electric field lines come into picture owing to Faraday's law [19]. The schematic presentation of the generation of circulating electric field is depicted in Fig. 4.11. The rotational electric field will induce a current inside the sample.

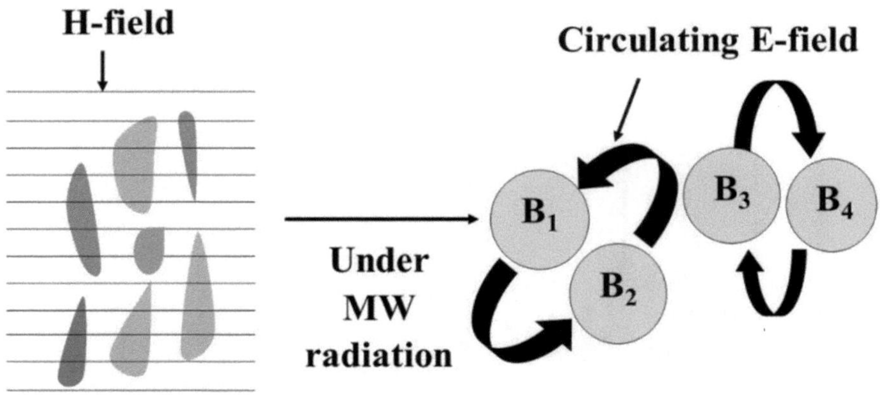

Fig. 4.11 A schematic presentation for the generation of circulating electric field

Thus, the interaction of H-field with constituent diamagnetic particles results in *circulatory atomic currents* (originated due to circulatory E-field as shown in Fig. 4.11), thereby making them behave like magnetic dipoles, which reorient in a direction opposite to the applied alternating magnetic field. This kind of *magnetic reorientation* leads to heating in the diamagnetic powders due to magnetic losses. Furthermore, the circulating electric current leads to an ohmic loss which results in joule heating of the sample. Thus, the heating of a diamagnetic semiconductive material is caused due to magnetic reorientation and ohmic losses.

4.5 Conclusion

The MW processing of semiconducting SiGe powder mixture in a 'single-mode' arrangement with domination of H-field has successfully resulted in the formation of $Si_{1-x}Ge_x$ alloys in different morphologies (nanostructures and nanowires) due to smaller particle size (less than 80 nm) in a short time (few minutes). It is the growth kinetics and higher reaction rates during MW processing of these powders which have resulted in the drastic reduction of alloying temperature (~400 °C). Both the smaller size and random shape of starting particles resulted in the MW interaction through the induction of magnetic losses. It has been possible to control morphology too by changing the ambient—argon or air—for the processing. The argon environment has resulted in the formation of polycrystalline $Si_{1-x}Ge_x$ ($0 \leq x \leq 0.25$) alloy. The processing in the air environment successfully resulted in the growth of extremely long (>10 μm) and randomly oriented nanowires. The oxygen, present in the air ambient, assisted the growth of nanowires. The enhanced growth of nanowires in the samples, containing higher concentration of germanium, is attributed to higher vapour pressure of germanium, as compared to silicon, at the growth temperature.

References

1. Duan, X., Huang, Y., Cui, Y., Wang, J., & Lieber, C. M. (2001). *Nature, 409*, 66.
2. Al-Jenabi, O., Aslam, Z., Cochrane, R. F., & Mullis, A. M. (2024). Microstructural characterisation and compound formation in rapidly solidified SiGe alloy. *Journal of Alloys and Compounds, 992*, 174560.
3. Douri, Y. A., Feng, Y. P., & Huan, A. C. H. (2008). *Solid State Communications, 148*, 521.
4. Noyan, I. D., Gadea, G., Salleras, M., Pacios, M., Calaza, C., Stranz, A., & Fonseca, L. (2019). SiGe nanowire arrays based thermoelectric microgenerator. *Nano Energy, 57*, 492–499.
5. Loharkar, P. K., Ingle, A., & Jhavar, S. (2019). Parametric review of microwave-based materials processing and its applications. *Journal of Materials Research and Technology, 8*(3), 3306–3326.
6. Ravindra, N. M., Sopori, B., Gokce, O. H., Cheng, S. X., Shenoy, A., Jin, L., Abedrabbo, S., Chen, W., & Zhang, Y. (2001). *International Journal of Thermophysics, 22*, 1593.
7. Dube, C. L., Kashyap, S. C., Dube, D. C., & Agarwal, D. K. (2009). Synthesis of silicon nanowires using microwave energy for optical devices. *International Journal of Green Nanotechnology: Materials Science & Engineering, 1*(1), M11–M15.
8. Dube, C. L., Kashyap, S. C., & Dube, D. C. (2015). Microwave growth mechanism of SiGe alloy nanowires synthesized in H-mode cylindrical cavity resonator. *Materials Discovery, 2*, 44.
9. Dube, C. L., Kashyap, S. C., Dube, D. C., & Agarwal, D. K. (2009). Room temperature photoluminescence and dc resistivity of Si0. 95Ge0. 05 alloy nanowires grown in microwave H-field. *Journal of Alloys and Compounds, 488*(1), 328–330.
10. Dube, C. L., Kashyap, S. C., Dube, D. C., & Agarwal, D. K. (2009). Growth of Si 0.75 Ge 0.25 alloy nanowires in a separated H-field by microwave processing. *Applied Physics Letters, 94*(21), 213107.
11. Lee, D. N., & Lee, S. B. (2015). Solid-phase crystallization of amorphous silicon films. In *Advanced Topics in Crystallization*. INTECH.
12. Cullity, B. D., & Smoluchowski, R. J. P. T. (1957). Elements of X-ray diffraction. *Physics Today, 10*(3), 50–50.
13. Kittel, C. (1976). *Introduction to solid state physics*. Wiley.
14. Mondal, A., Agrawal, D., & Upadhyaya, A. (2010). Microwave sintering of refractory metals/ alloys: W, Mo, Re, W-Cu, W-Ni-Cu and W-Ni-Fe alloys. *Journal of Microwave Power and Electromagnetic Energy, 44*(1), 28–44.
15. Thompson, K., Booske, J. H., Ives, R. L., Lohr, J., Gorelov, Y., & Kajiwara, K. (2005). Millisecond microwave annealing: Driving microelectronics nano. *Journal of Vacuum Science & Technology B: Microelectronics and Nanometer Structures Processing, Measurement, and Phenomena, 23*(3), 970–978.
16. Brodovoi, A. V., Bunchuk, S. G., Pokropivny, V. V., & Skorokhod, V. V. (2006). Magnetic properties of nanoporous Si powder. *International Journal of Nanotechnology, 3*(1), 57–62.
17. Ryabakov, K. I., Semenov, V. E., Egorov, S. V., Eremeev, A. G., Plotnikov, I. V., & Bykov, Y. V. (2006). *Journal of Applied Physics, 99*, 023506.
18. Cullity, B. D., & Graham, C. D. (2011). *Introduction to magnetic materials*. John Wiley & Sons.
19. Lomon, E. L. (1962). In J. D. Jackson (Ed.), *Classical electrodynamics* (p. xvii+ 641 Illus. $13). Wiley.

Chapter 5
Microwave Processing of Dilute Magnetic Semiconductors

Abstract This chapter delves into MW processing of spintronic materials, specifically dilute magnetic semiconductors (DMSs). A brief description of DMS is followed by the MW-assisted synthesis of dilute magnetic semiconductors which include both the group IV semiconductors and groups II–VI compound semiconductors. These are polycrystalline Mn-doped SiGe alloys and polycrystalline ferromagnetic Zn1-xCoxOy. This processing technique has demonstrated the extension of equilibrium solubility of Mn, by several orders, in SiGe alloy. A detailed structural and magnetic characterization of both the systems is given in this chapter.

5.1 Dilute Magnetic Semiconductors

In recent years, doped semiconductors, both elemental and compound, in which a small doping of magnetic atoms/ions (transition metal ions, i.e. Co, Mn, etc.) is added in the host semiconductor, have potential application primarily because these doped semiconductors exhibit ferromagnetism at room temperature and therefore are employed in a wide range of applications in spintronics. Broadly, these semiconductors displaying/exhibiting ferromagnetism at room temperature (RTFM) are termed as dilute magnetic semiconductors (DMSs).

The Si-based transition metal (TM)-doped DMSs can, in fact, be a best suited material for spintronics applications as the spintronics devices can be integrated on one and the same chip. To get lattice matched and with a possibility of band gap engineering, SiGe alloy is preferred over a Si/Ge elemental matrix. The solid solubility of Mn in Si and Ge is very low (~10^{16} cm^{-3} at 1000 °C) [1]. In a binary system, the solubility is predicted by Hume-Rothery rules [2]. The low solid solubility of TM element in Si and Ge necessitates the use of nonequilibrium growth techniques such as ion implantation which can incorporate dopants into semiconductors beyond the solid solubility limit [3]. The Mn can dissolve up to certain limits only into the Si structure at a given temperature using conventional heating methods. This is because of factors like the difference in atomic size and electronegativity between Mn and Si, which can hinder their ability to form a discontinuous solid

C. L. Dube, *Microwave Assisted Synthesis of Materials*, SpringerBriefs in Materials, https://doi.org/10.1007/978-3-032-01439-9_5

solution [4, 5]. MW processing is a nonequilibrium method; the presence of MW fields assists in extended solubility of dopant in the host matrix [6]. An anisothermal situation during MW processing promotes nonequilibrium reaction that creates a temperature difference between different constituents, thus inducing a unique driving force for a solid–solid diffusion and providing ways for nonequilibrium reaction pathways [7].

In the case of compound semiconductors/oxides, the theoretically stable ferromagnetic state in TM-doped ZnO is predicted, which has a potential application as a magnetic nanomaterial for spintronics [8]. In some cases, the ferromagnetic ordering was attributed to the congregating of TM elements and to the formation of secondary phases. Different methods were employed to synthesize TM element doped ZnO and it is observed that induction of ferromagnetic ordering in ZnO is sensitive to the synthesis method and morphology [9].

5.2 Synthesis of Dilute Magnetic Semiconductors

A novel and a rapid nonequilibrium microwave-assisted method for material synthesis has been employed for the preparation of both pristine (undoped) and doped tetrahedrally coordinated semiconductors (SiGe:Mn) and oxide-semiconductors (ZnO:Co) [10]. It is believed that the interaction of MW with the starting precursors of the material at the atomic level can induce RTFM in SiGe:Mn and ZnO:Co based DMSs. The synthesis and characterization of ZnO:Co will follow that of SiGe:Mn as described in the following subsections.

The starting precursors, to synthesize samples of different compositions of $Si_{0.95}Ge_{0.05}$: Mn (Mn content of 2, 5, and 7 weight %) and ZnO:Co (Co content of 5 weight %), were taken desired proportions and mixed well. In each case, the resultant mixture was pressed into a pellet and exposed to H-field of MW radiation (in a single-mode applicator). The pellets of composition of $Si_{0.95}Ge_{0.05}$: Mn were processed in both air and Ar ambience, while different compositions of ZnO:Co were processed only in air ambience. The MW power input was adjusted to obtain desirable temperature. The temperature acquired by the sample was determined using the pyrometer set at an appropriate emissivity value (0.6 for SiGe and 0.15 for ZnO-based samples). The pellets of $Si_{0.95}Ge_{0.05}$: Mn were labelled as 5SM0 (0 weight % Mn), 5SM2 (2 weight % Mn), 5SM5 (5 weight % Mn), and 5SM7 (7 weight % Mn) and processed at 900 °C for 5 min in air and Ar ambience separately. The $Zn_{0.95}Co_{0.05}O$ pellets were processed in the air ambience only at 1000 °C for 10, 20, 40, and 60 min.

The structure and phase formation of MW processed samples were analysed by employing X-ray diffractometer. Magnetization studies were carried out using a vibrating sample magnetometer. The microstructures of the synthesized samples were investigated by using a scanning electron microscope [10].

5.3 Characterization of Manganese-Doped SiGe Alloy

5.3.1 Structural Analysis

Figure 5.1 shows the X-ray diffraction patterns of 5SM0, 5SM2, 5SM5, and 5SM7 samples. The XRD patterns reveal that all the samples (except 5SM7) are phase pure exhibiting diamond structure of SiGe alloy. This indicates that with possible incorporation of Mn in SiGe alloy, a substitutional alloy forms, and there is no change in the lattice structure.

The XRD patterns for 5SM0 and 5SM5 samples are replotted in part (between diffraction angle 76 and 77 degree) in Fig. 5.2. It is noted that the (331) peak corresponding to $Si_{0.95}Ge_{0.05}$: 5 wt % Mn is shifted towards lower diffraction angle in comparison to that of pristine $Si_{0.95}Ge_{0.05}$ sample. The XRD peak position depends on the lattice constant. The Mn doping causes the crystal lattice to expand, resulting in a larger d-spacing and a smaller diffraction angle according to Bragg's Law. This is understandable as the ionic radii of Mn^{4+}, Si^{4+}, and Ge^{4+} in tetrahedral coordination are 0.39, 0.26, and 0.39 Å, respectively [11]. The observed shift towards lower diffraction angle is attributed to the fact that ionic radii of Mn ion in any oxidation state are larger than that of silicon and germanium radii [11]. This finding confirms that Mn is substitutionally accommodated in $Si_{0.95}Ge_{0.05}$ lattice.

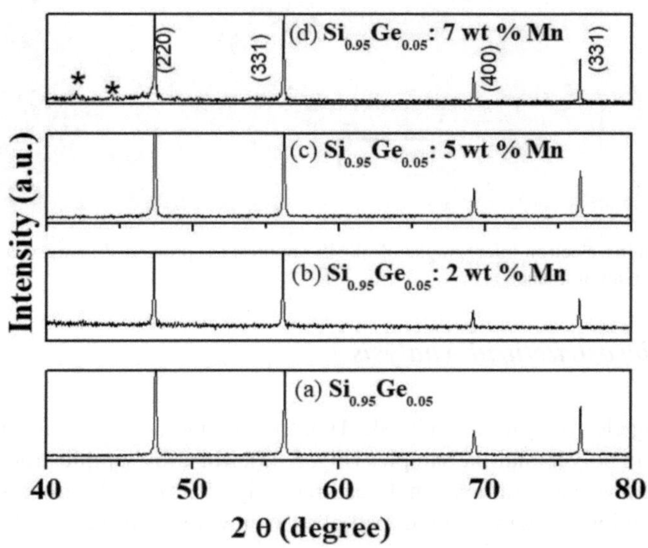

Fig. 5.1 The XRD patterns of (**a**) pure SiGe, (**b**) $Si_{0.95}Ge_{0.05}$: 2 wt % Mn, (**c**) $Si_{0.95}Ge_{0.05}$: 5 wt % Mn, (**d**) $Si_{0.95}Ge_{0.05}$: 7 wt % Mn samples

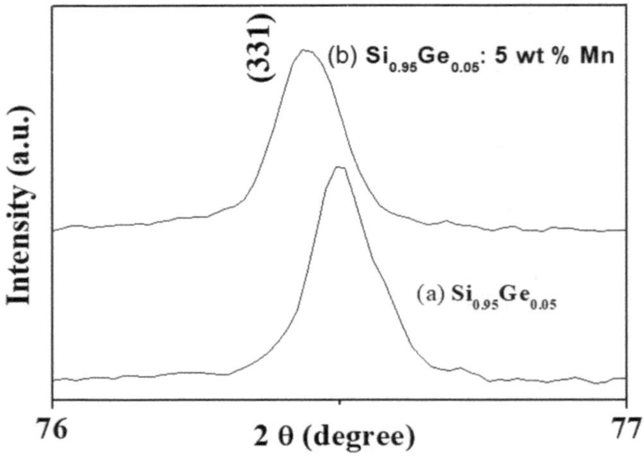

Fig. 5.2 XRD patterns [(331) peak only] of (**a**) undoped SiGe alloy and (**b**) $Si_{0.95}Ge_{0.05}$: 5 wt % Mn alloy samples

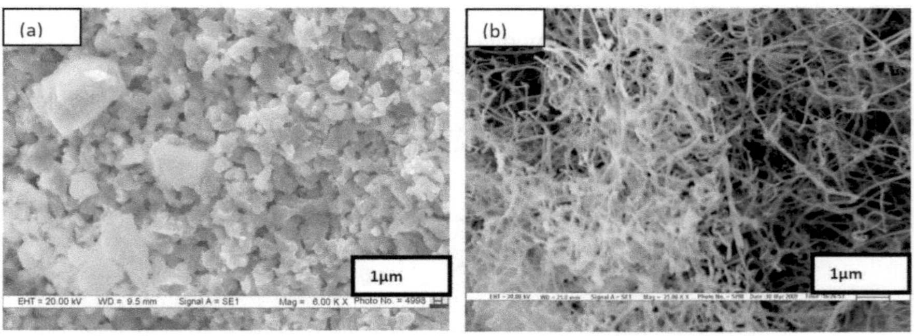

Fig. 5.3 Scanning electron micrographs of 5SM5 samples: (**a**) processed in argon ambience and (**b**) processed in air ambience

5.3.2 Microstructural Analysis

The scanning electron micrograph (SEM) of the $Si_{0.95}Ge_{0.05}$: 5 wt % in Ar ambience has resulted in polycrystalline samples (Fig. 5.3a). Mn alloy sample processed in air ambience shows nano-dimensional features (Fig. 5.3b). MW processing in air shows the random growth of the doped alloy nanowires with very high yield. The grown nanowires in air ambient are of nearly uniform diameter (~100 nm). Most of the nanowires are extremely long (>10 μm) and randomly oriented.

5.3.3 Magnetic Characterization

The magnetization vs. field (M-H) loops for MW-processed (Ar ambience) Mn-doped SiGe alloy samples: 5SM2, 5SM5, and 5SM7 are shown in Fig. 5.4a. The M-H loops recorded at room temperature, therefore, clearly indicate the induction of ferromagnetic ordering in $Si_{0.95}Ge_{0.05}$: Mn system. The M-H loops indicate the increase in ferromagnetic ordering up to 5 wt % of Mn in $Si_{0.95}Ge_{0.05}$ system with saturation magnetization value ~450 emu/gm and coercivity value~3mT. Further, the continuous increase in magnetic moment of the MW processed samples with increase in applied field possibly indicates the presence of paramagnetic phases along with ferromagnetic ordering in the sample. It is attributed to the fact that all magnetic phases of binary SiMn and GeMn alloys have been reported to have their Curie temperature below room temperature.

Thus, the ferromagnetism with Curie temperature over 300 K has been observed in polycrystalline SiGe:Mn alloy samples processed in Ar. That Mn is successfully incorporated in the SiGe host lattice of diamond structure in these samples is confirmed by the XRD patterns. The results indicate that Mn may be substitutionally added at concentrations of several orders of magnitude greater than its equilibrium solubility limit, possibly due to extended solid solubility induced by MW interaction with host semiconducting matrix occurring at atomic level [12]. It is reported the use of nonequilibrium growth techniques, such as ion implantation and MW processing, can introduce dopants into semiconductors beyond the solid solubility limit [3]. Magnetization decreases with an increase in Mn-concentration beyond 5 wt. % of Mn. It is also to be noted that secondary phases are found at higher concentrations of the magnetic dopant and the same has been attributed to the

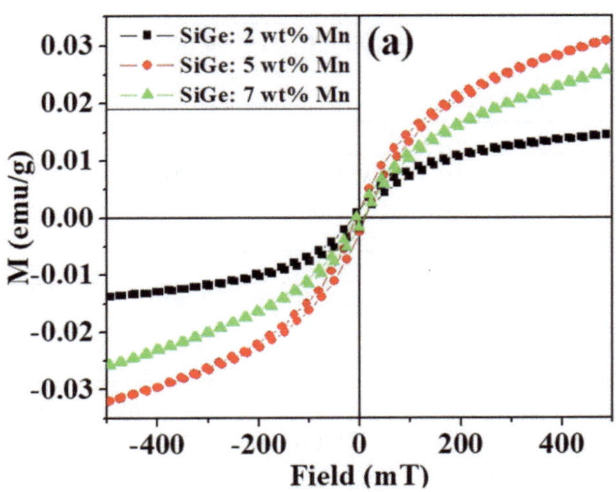

Fig. 5.4 (**a**) As obtained M-H loops for MW processed Mn-doped SiGe alloy (5SM2, 5SM5, and 5SM7) samples

weakening of ferromagnetic interaction since the ferromagnetism is usually observed at lower concentrations of the magnetic dopant.

5.4 Characterization of Cobalt-Doped ZnO

5.4.1 *Structural Analysis*

The XRD patterns of the samples synthesized at different experimental conditions are shown in Fig. 5.5. These XRD patterns by and large reveal the wurtzite structure of ZnO. The reflection corresponding to (002) planes of Co, present in the XRD of the unprocessed sample, is absent in the XRD patterns of MW processed samples, thus indicating the possible incorporation of most of Co into ZnO lattice.

The XRD patterns of the MW processed samples processed in different experimental conditions are shown in Fig. 5.5.

The XRD patterns (in zoomed-in scale) of the starting ZnO and Co mixture and of MW processed samples in air ambient at 1000 °C for 60 min are shown in Fig. 5.6.

The low intensity reflections marked by/with * (Fig. 5.6) correspond to CoO as a minor phase present in a sample. The formation of cobalt oxide (CoO) is

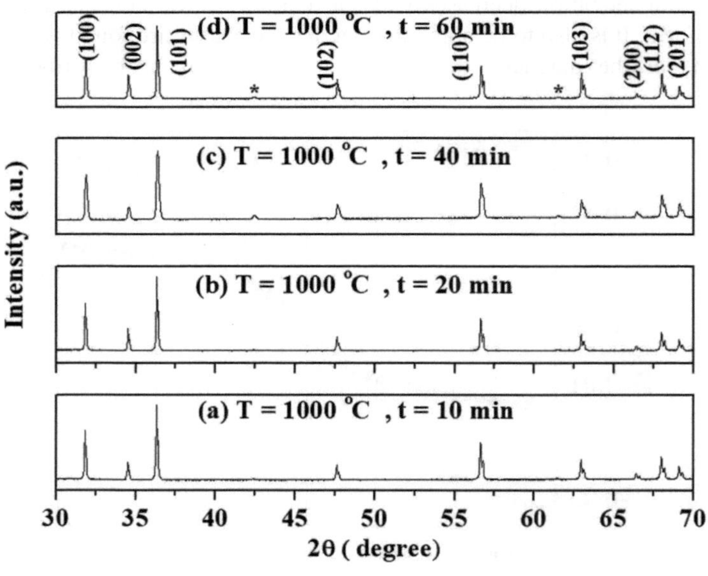

Fig. 5.5 XRD patterns of H-field processed in air ambience ZnO:Co mixture at 1000 °C for (**a**) 10 min, (**b**) 20 min, (**c**) 40 min, and (**d**) 60 min

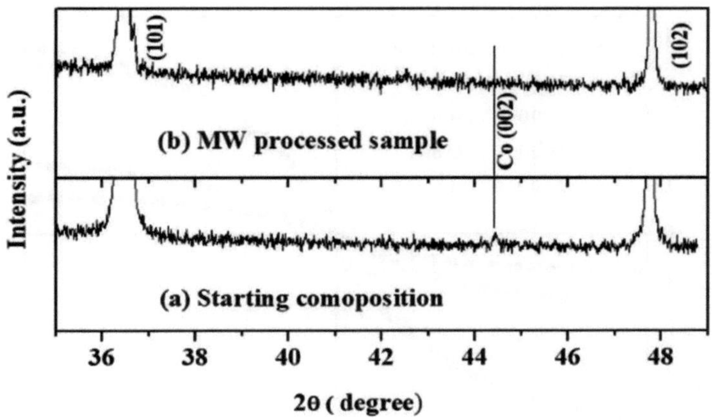

Fig. 5.6 XRD patterns of (**a**) starting mixture and (**b**) H-field processed ZnO:Co sample at 1000 °C for 60 min

understandable as the MW processing is performed in air ambience at a temperature of 1000 °C for 60 min. [13]. A small change in lattice parameters is understandable. The lattice parameter *a* increases (3.23–3.24 Å) slightly, whereas the lattice parameter *c* decreases (5.20–5.19 Å) by a nominal value over the respective lattice parameter values of the pure ZnO phase, thereby leaving the volume of unit cell practically unchanged. The observed small change in lattice parameters is due to the substitution of Co^{2+} (0.58 Å) for Zn^{2+} (0.60 Å) at the tetrahedral coordination in ZnO lattice. Thus, XRD data is direct evidence for Co substituting Zn into ZnO lattice as Co^{2+} at the tetrahedral coordination. Further, X-ray photoelectron spectrum constituent elements in H-field processed ZnO:Co sample at 1000 °C for 60 min confirms the incorporation of Zn^{+2} with *high spin* Co^{+2} in the ZnO lattice [14].

5.4.2 Magnetic Characterization

The M-H loops of the MW samples processed for 10, 20, 40, and 60 min in air ambience are shown in Fig. 5.7. The decrease in magnetic moment of the MW processed samples with increase in processing time is due to more and more cobalt substituting Zn and resulting in weak ferromagnetic $Zn_{1-x}Co_xO_y$, where x is less than 0.05 wt.%.

A room temperature M-H loop for the MW synthesized (60 min at the temperature of 1000 °C) exhibiting ferromagnetic order is shown in Fig. 5.8. This is obtained from the raw data after subtracting paramagnetic contribution from the observed magnetic moment [10].

The saturation magnetization and coercivity of the sample processed at 1000 °C in for 60 min are found to be 0.6 emu/g and 4 mT, respectively.

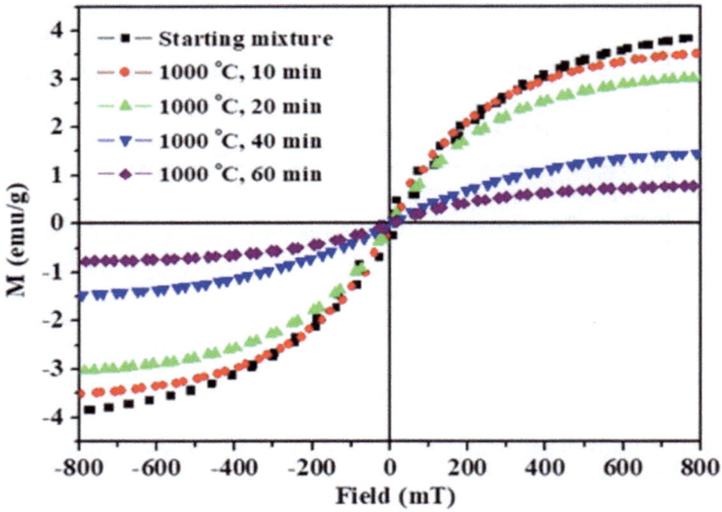

Fig. 5.7 As obtained M-H loops for starting mixture and MW processed samples for 10, 20, 40, and 60 min at the temperature of 1000 °C

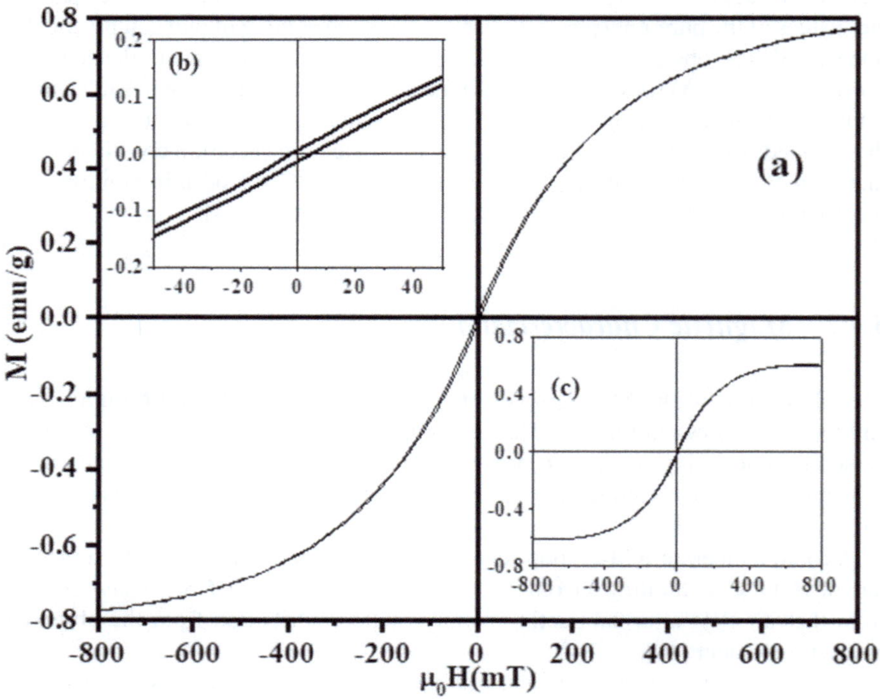

Fig. 5.8 The RT M-H loop of the MW-processed $Zn_{0.95}Co_{0.05}O$ sample: (**a**) As-obtained raw data. (**b**) The M-H loop replotted, showing small magnetization and coercivity. (**c**) The ferromagnetic fraction withdrawn from the raw data

5.5 Conclusions

Microwave processing, in a separated H-field, has successfully been employed to obtain ferromagnetic manganese-doped SiGe alloys at 900 °C in 5 min in air as well as Ar ambience. The observed change in lattice parameter, as estimated by X-ray diffractometry, has established the substitution of Mn in SiGe alloy. The M-H loops, recorded at room temperature, clearly indicate the induction of ferromagnetic ordering in $Si_{0.95}Ge_{0.05}$: Mn system. Likewise, polycrystalline ferromagnetic bulk $Zn_{1-x}Co_xO_y$ samples have also been synthesized in air ambiance at the temperature of 1000 °C in 60 min. The X-ray diffractogram has established the wurtzite structure of the Co-doped ZnO samples with the understandable modified lattice parameters. The room temperature magnetization study validates the ferromagnetic ordering in the microwave-processed samples.

References

1. Bolduc, M., Awo-Affouda, C., Stollenwerk, A., Huang, M. B., Ramos, F. G., Agnello, G., & LaBella, V. P. (2005). Above room temperature ferromagnetism in Mn-ion implanted Si. *Physical Review B, 71*(3), 033302.
2. Callister, W. (2007). *4.3 impurities in solids, in: Materials science and engineering and introduction* (7th ed., pp. 83–85). John Wiley and Sons.
3. Liu, F., Prucnal, S., Berencén, Y., Zhang, Z., Yuan, Y., Liu, Y., & Zhou, S. (2017). Realizing the insulator-to-metal transition in Se-hyperdoped Si via non-equilibrium material processing. *Journal of Physics D: Applied Physics, 50*(41), 415102.
4. Nordberg, G. F., Fowler, B. A., & Nordberg, M. The handbook on the toxicology of Metalsis a comprehensive review dealing with the effects of metallic elements and their compounds on biological systems. In *Special emphasis has been laid on the toxic effects in humans, although toxic effects in animals and biological systems in vitro are*.
5. Ohno, H. (1998). Making nonmagnetic semiconductors ferromagnetic. *Science, 281*(5379), 951–956.
6. Phan, P. T., Hong, J., Tran, N., & Le, T. H. (2023). The properties of microwave-assisted synthesis of metal–organic frameworks and their applications. *Nanomaterials, 13*(2), 352.
7. Fukushima, J. (2024). Synthesis of functional ceramics by microwave non-equilibrium processing. *Journal of the Ceramic Society of Japan, 132*(3), 79–84.
8. Sato, K., & Katayama-Yoshida, H. (2001). Stabilization of ferromagnetic states by electron doping in Fe-, Co-or Ni-doped ZnO. *Japanese Journal of Applied Physics, 40*(4A), L334.
9. Bhatti, K. P., Chaudhary, S., Pandya, D. K., & Kashyap, S. C. (2006). Extrinsic nature of the room temperature ferromagnetism in (ZnO) 1− x (MnO2) x for 0.01≤ x≤ 0.97. *Solid State Communications, 140*(1), 23–27.
10. Dube, C. L., Kashyap, S. C., Dube, D. C., & Agarwal, D. K. (2008). Microwave processing of ZnO based dilute magnetic semiconductors. In *2008 international conference on recent advances in microwave theory and applications* (pp. 126–129). IEEE.
11. Perry, J. H. (1941). *Chemical engineers' handbook*. McGraw-Hill.
12. Zhang, Y., Jiang, Q., Smith, D. J., & Drucker, J. (2005). Growth and characterization of $Si_{1-x}Mn_x$ alloys on Si (100). *Journal of Applied Physics, 98*(3), 033512.

13. Dube, C. L., Kashyap, S. C., Dube, D. C., & Agarwal, D. K. (2008, November). Microwave processing of ZnO based dilute magnetic semiconductors. In *2008 international conference on recent advances in microwave theory and applications* (pp. 126–129). IEEE.
14. Kashyap, S. C., Dube, C. L., & Kotnala, R. K. (2009, December). Rapid decrystallization of a polycrystalline hard ferrite by separated H-field microwave processing. In *2009 applied electromagnetics conference (AEMC)* (pp. 1–1). IEEE.

Chapter 6
Microwave Processing of Technologically Important Materials

Abstract The MW-assisted synthesis of materials is a recent and a novel method. This chapter discusses the microwave-assisted synthesis of technologically important materials, such as functionally graded (FG) tungsten copper for future reactors, nanocrystalline ferrite (Fe_3O_4) for targeted drug delivery, magnetic nanocomposites for groundwater remediation, microwave-assisted synthesis of glass for NIR shielding, and glass ceramic for nuclear waste immobilization. The H-field of microwave radiation at microwave frequency (2.45 GHz) is utilized for phase transformation/ decrystallization of M-type hard barium hexaferrite (from crystalline to amorphous). The decrystallized hexaferrites are found to be good alternative RE-free permanent magnet.

6.1 Introduction

In the previous chapters, the MW-assisted synthesis of different kinds of materials was discussed. The samples include group IV semiconducting alloys, in both virgin and transition metal-doped forms along with dilute magnetic II–VI semiconductors (spintronic materials). These materials find applications in several fields. This chapter discusses the microwave-assisted synthesis of metallic materials, such as functionally graded (FG) tungsten copper for future reactors, nanocrystalline ferrite (Fe_3O_4) for targeted drug delivery, magnetic nanocomposites for groundwater remediation, microwave-assisted synthesis of glass for NIR shielding, and glass ceramic for nuclear waste immobilization. The electromagnetic energy (associated with H-field) at microwave frequency (2.45 GHz) is utilized for phase transformation/decrystallization of M-type hard barium hexaferrite (from crystalline to amorphous). The decrystallized hexaferrites are having low magnetocrystalline anisotropy and can prove to be a good alternative to RE-free permanent magnet.

C. L. Dube, *Microwave Assisted Synthesis of Materials*, SpringerBriefs in Materials, https://doi.org/10.1007/978-3-032-01439-9_6

6.2 Functionally Graded (FG) Tungsten Copper for Future Reactors

A functionally graded material (FGM) is often utilized for the removal of hurdles that arise from the joining of two unlike metals which have different mechanical properties [1]. A bimetallic functionally graded compact of tungsten copper (W/Cu) has a unique application in future fusion devices [2]. The plasma facing components (PFC) in the divertor of an ITER-like tokamak receive high heat loads and are made up of mainly tungsten/tungsten alloys and a copper-chromium-zirconium alloy to remove the heat from divertor. But joining of tungsten to copper is a daunting task due to huge mismatch of *coefficients of thermal expansion* (α) of the tungsten and alloy [3]. To bring down the mismatch between tungsten and alloy, W/Cu FGM can be used as a *sandwich layer* between tungsten/tungsten alloys and the copper alloy. The purpose of the incorporation of *sandwich layer* is to mitigate undesirable cracking, delamination, etc.

There are many methods for the synthesis of FGMs, such as powder metallurgical, spark plasma sintering, etc. [4].

The first report on MW processing of metals was published in 1999 [5] and was followed by several researchers [6–8]. The W/Cu FGM can be prepared by microwave (MW) heating method [9]. The skin depth of microwaves in metals at room temperature is typically in the range of 1–5 μm [6]. During MW heating, volumetric heating takes place via a loss mechanism as discussed in Chap. 3. The MW method results in fine microstructure, hence improved mechanical properties [10]. The finer microstructure is retained due to rapid heating rate and reduced processing time, both of which hinder the grain growth. The samples with finer microstructure exhibit enhanced irradiation resistance as well, which is desired for plasma facing components in future fusion devices [11]. It may be pointed out that the MW processing had successfully been employed for the synthesis of bulk tungsten, copper, and W/Cu composites [1]. Invariably, each individual metallic powder particle in metallic compact is encapsulated in a dielectric oxide layer, which prevents *connectivity percolation* and thereby enhanced absorption of microwaves [7]. In addition, the skin/penetration depth of MW radiation in metal plays a vital role in heating metallic powder compacts. The skin depth for tungsten and copper at 2.45GHz is in the micron regime [7]. Due to skin depth in the micrometre regime, bigger metallic particles (more than *skin depth of MW* in respective metal) cannot effectively absorb microwaves at room temperature. To vanquish this problem in poor absorbing materials, a *hybrid heating (HH) method* is being used. The HH method utilizes both MW besides conventional heating along with a ceramic susceptor (SiC) which is invariably used to first raise the temperature of a specimen up to a desired level [1]. Another HH method, such as *plasma-assisted microwave heating*, does not require any ceramic susceptor. The tungsten/copper FGM was successfully fabricated at a temperature of 800 °C and in a multimode MW applicator in 30 min [1]. The microstructure and elemental composition of a W/Cu FGM specimen, fabricated by this method, had been analysed by a scanning electron microscope. The scanning electron micrograph (SEM) constructed from *backscattered electrons* (*BSEs*) of the sample is shown in Fig. 6.1. The tungsten has higher atomic number (74) than copper (29) and hence will have greater chances for an *elastic collision* with incident electrons. The number of BSEs emitted from tungsten will be higher, as compared to copper, thereby brighter region in BSE image corresponds to tungsten in the sample. The quasi-homogeneous variation in tungsten content along the length of compact can be seen in the inset of Fig. 6.1. The change of tungsten/copper content along the radial direction is nearly consistent, and the variation of elements along the length of the FGM confirms that the synthesized FGM sample is one-dimensional.

The Vickers hardness value and copper distribution across the FGM sample are shown in Fig. 6.2. The hardness of the FGM sample increases with the increase in tungsten amount. The increase in hardness is intelligible due to the hardness value of tungsten as compared to copper. It can also be seen that indentation size decreases with the increase in W content along the direction (shown by arrow) in the inset image of Fig. 6.1 due to an increase in the hardness of the sample.

The SEM image and EDAX analysis of tungsten and copper across the length of compact reveal formation of the graded structure [1]. Low temperature synthesis of FGM samples is attributed to enhanced diffusion kinetics owing to *non-thermal microwave effects*. The MW-assisted synthesis assisted in achieving the fine structure of FGM sample and thereby resulted in good mechanical property.

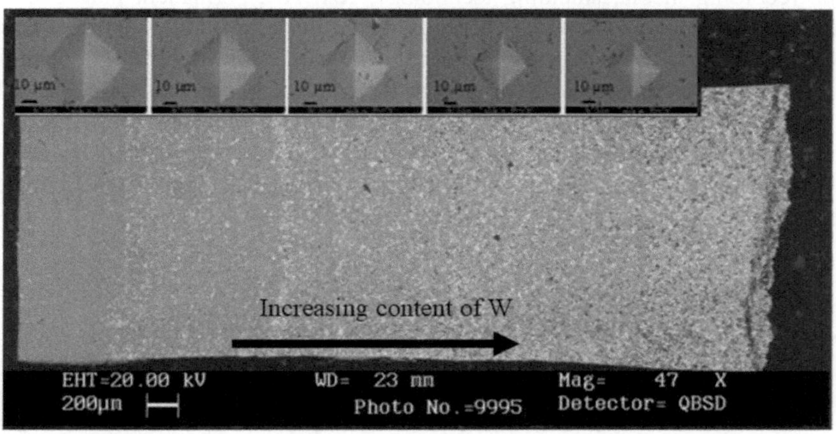

Fig. 6.1 SEM of FGM sample and indentations made along the length of the FGM

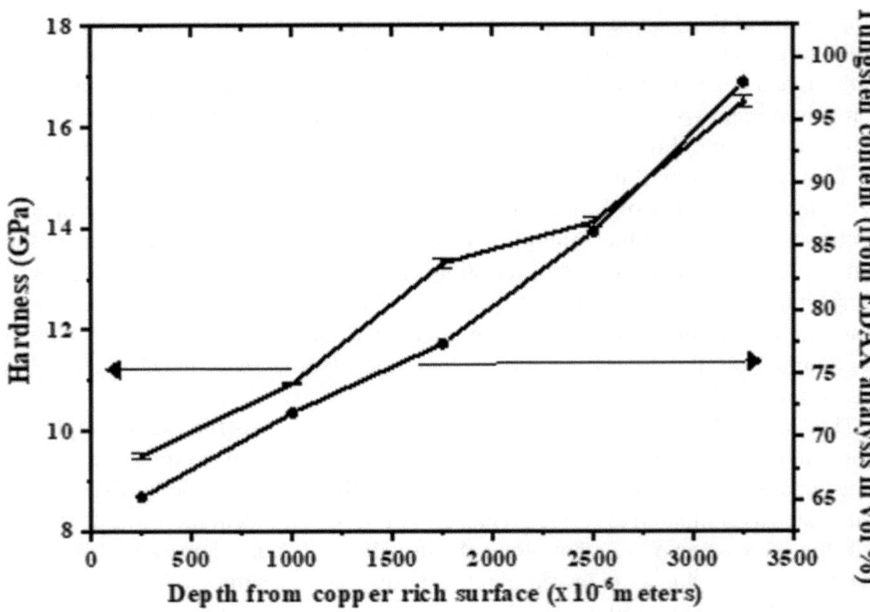

Fig. 6.2 Vickers hardness and copper distribution along the depth of W/Cu FGM sample

6.3 Nanocrystalline Ferrite for Diverse Applications

Nanocrystalline ferrite finds application in various fields, such as spintronics, magnetic recording media, shielding technology, RADAR, permanent magnets, and microwave attenuators [12]. The efforts are being made to synthesize theses nanocrystalline ferrites by various methods [13]. The nanoferrites have found potential applications in the biomedical field too as their magnetic properties can easily be controlled by an external magnetic field. The tuning of magnetic properties with external magnetic fields makes them potential nanomagnets in various usage like targeted drug delivery, magnetic separation, biosensors, MRI, and magnetic hyperthermia [14].

6.3.1 Decrystallized Hard Ferrites for Rare-Earth (RE) Element-Free Permanent Magnet

The microwave processing of materials may result in phase transformation from polycrystalline to decrystallized/nanocrystalline phase. Many studies have been made on MW processing of ceramics, such as ferrites [15–17], alumina [18],

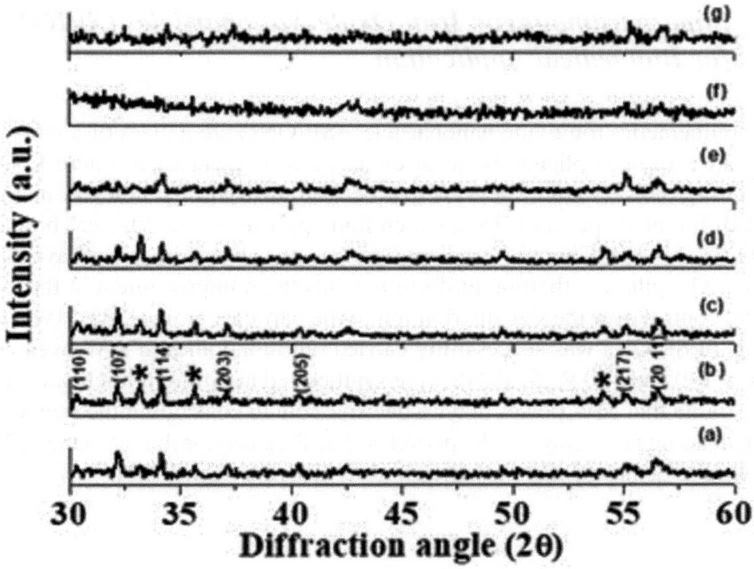

Fig. 6.3 XRD patterns of barium hexaferrite samples conventionally prepared in a muffle furnace at (**a**) 1000 °C, 5 h and MW-assisted phase transition processed at (**b**) 600 °C, 10 min, (**c**) 800 °C, 10 min, (**d**) 900°C, 5 min, (**e**) 900 °C, 10 min, and (**g**) 1000 °C, 2 min

carbide [19], and composites materials like WC/Co [20]. The demand for high-performance permanent magnets is growing expeditiously, because of their soaring applications in diverse industries. A high-performance permanent magnet contains rare-earth (RE) elements. However, the rare earth (RE)-based permanent magnets are exorbitantly high in price with limited supply. Hence, the venture to produce alternative RE-free permanent magnets is getting more and more attention [21]. In an effort, the M-type barium hexaferrite ($BaFe_{12}O_{19}$) was processed at frequency of 2.45 GHz, in a temperature range of 600–1000 °C and for different durations (2–10 min) [17]. The X-ray diffractograms of MW processed M-type hexaferrites (Fig. 6.3) confirm complete phase transformation, from crystalline to decrystallization, of the MW processed sample [17] with increase in process parameters (both temperature and duration). The high value of magnetostriction constant is responsible for the decrystallization of the ferrites, processed in a multimode MW system at 28 GHz [22]. The VSM measurement indicates that the specific magnetization value for conventionally calcined sample is found to be 35 emu/g at a field of 480 mT and is increased to 52 emu/g for an H-field processed sample [17].

6.3.2 *Superparamagnetic Iron Oxide Nanoparticles (SPIONs) for Biomedical Application*

Superparamagnetic iron oxide nanoparticles (SPIONs) are a class of nanoparticles which find various applications in different areas of biomedical interest, such as targeted drug delivery, etc. The MW can be well utilized to synthesize nanostructures of different shapes and sizes, which find applications in different biomedical fields [1, 7, 11–15]. Thermal gradients are mostly absent in microwave-assisted synthesis. The inherent thermal uniformity leads to an improvement in the crystallinity and control over the size distribution of the particles/grains [23]. MW-assisted synthesis of SPIONs was successfully carried out in a domestic MW oven at MW powers of 180 and 450 W. SPIONS were synthesized at different microwave power, and it is found that MW power plays a decisive role in controlling the size, coating, and drug loading percentage of the particles. The drug (doxorubicin) loaded SPIONs were found to be promising and efficient drug delivery agents [24].

6.3.3 *Microwave-Assisted Synthesis of Magnetic Nanocomposites for Groundwater Remediation*

The groundwater contamination by heavy metal ions is detrimental to human health and also to flora and fauna. The heavy metal ions have a relatively high density (3.5–7 g/cm^3) against water 1 g/cm^3 and are extremely toxic even in low concentrations (0.1 mg/L) [25]. The heavy metal ions are toxic to the kidneys, nervous and reproductive system, etc. Owing to the low concentration density of toxic metal ions in water, it is very difficult to detect and remove the toxic elements using existing technology. The magnetic nanoparticles can be utilized in the field of groundwater remediation for the removal of heavy metal ions through the adsorption method [26]. The bare magnetite nanoparticles form aggregates in air and hence lead to poor adsorption capacity [26]. To enhance the adsorption capacity, magnetic nanoparticles can be functionalized with a good adsorbent, such as graphene oxide (GO); the resulting magnetic graphene oxide (M/GO) is preferred over other existing materials due to its high surface area to volume ratio and magnetic nature [27]. The higher the surface area, the larger the more active binding sites, which facilitate heavy metal ions to enter easily into interlayer spacing and form particles [28].

The multi-layer graphene oxide (GO) and graphene oxide-based magnetic nanocomposites were synthesized by microwave-assisted modified Hummers' method [27, 29]. The experimental details of the synthesis and adsorption studies are discussed elsewhere [27]. The high-resolution TEM image of graphene oxide-based magnetic nanocomposite is shown in Fig. 6.4.

The spherically shaped Fe_3O_4 crystals were found to be smoothly distributed on the GO sheets, as shown in Fig. 6.4. The selected area electron diffraction pattern showed diffraction spots in a randomly dotted ring pattern, indicating that the GO/

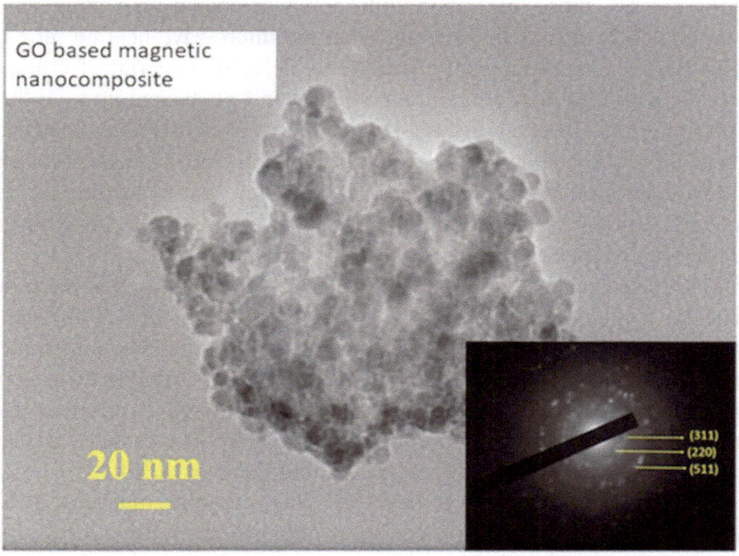

Fig. 6.4 The high-resolution TEM image of graphene oxide-based magnetic nanocomposite

Fe_3O_4 nanocomposite is of a polycrystalline nature and well indexed to the cubic (fcc) phase Fe_3O_4 structure [29]. The GO/Fe_3O_4 nanocomposite have potential application in the filtration of wastewater containing heavy metal ions in the low and intermediate concentration range [29].

6.4 Microwave-Assisted Synthesis of Glass and Glass Ceramic

MW-assisted synthesis of glasses has gained attention due to the energy efficiency of synthesis method, energy savings (i.e. economic), and ability to achieve uniform heating during MW synthesis [30]. The microwave radiation is employed to heat/melt raw materials (mainly oxides), offering several advantages over conventional furnace-based melting processes. The method is relatively a recent development for the synthesis of bulk glass at an industrial scale. An expeditious and uniform heating throughout the material is achieved due to the volumetric heating of MW radiation absorption [31]. MW heating is material-dependent heating. It can preferentially heat certain components, leading to enhanced reaction kinetics. The MW-assisted synthesis method has the ability to tailor glass properties by adjusting MW parameters having improved hardness and enhanced conductivity (as compared to conventionally synthesized glasses).

The crystallization behaviour in glasses can be widely explored using a variable frequencies MW-assisted method. Attempts have been made to utilize MW energy for making glasses, those reinforced with metals or coated with films, as well as glass ceramics [32]. It may be reiterated that the microwave heating relies on the interaction between electromagnetic waves and dielectric materials. Certain dielectric materials with high loss couple immediately and get heated up to 1000 K in a very short time [33]. However, in case of poor microwave absorbing dielectric material such as SiO_2, pre-heating of the precursor with the help of susceptors assists in strong MW absorption in silicate glasses [31]. The microwave heating has recently been utilized as a tool for the modulation of copper redox ratio to improve the fluorescence activity of the zinc-borate glass [34].

Different glass systems comprising SiO_2, B_2O_3, and P_2O_5 as the main building block can be synthesized between the temperature range of 1200–1500 °C by MW-assisted synthesis method. The MW-assisted synthesis of non-stoichiometric tungsten oxide-doped borosilicate glasses has resulted in samples with good NIR shielding property characteristics. These samples exhibited reduced oxidation of tungsten state as compared to conventionally prepared. The reduced oxidation of tungsten is found responsible for good NIR shielding ability [35].

Microwave-assisted (MW) synthesis of glass-ceramic matrices is a novel approach that leverages microwave energy to enhance the melting, crystallization, and sintering processes involved in glass-ceramic production [36]. In this method, a parent glass composition is synthesized by melting oxides, quenching, and then grinding into fine powder. Subsequently, nucleating agents like TiO_2 or ZnO are added to control crystallization, and the mixture is exposed to MW radiation, leading to a controlled crystallization. After the natural cooling, the final glass-ceramic matrix is obtained. The MW-assisted hydrothermal synthesis of ZnO nanoparticles with high defect concentration is synthesized using microwave energy. An *accelerated crystallization* was found responsible for enhanced visible light photocatalysis [37].

The synthesis of bioglass ceramics with good mechanical and antibacterial properties for biomedical applications has been reported [36]. The MW-assisted one-step method can prepare tailing/tailored glass ceramics (TGC) with the help of the MW exposure at/of different parts of TGC utilizing different microwave susceptors, such as SiC graphite, active carbon, etc. [38]. Glass and ceramic precursors absorb MW radiation efficiently and lead to rapid heating. This technique significantly reduces the duration of processing, sintering, and/or crystallization as compared to the time taken during conventional furnace heating.

The MW-assisted synthesis method can be employed for glasses for nuclear waste immobilization. The most added advantage of microwave-assisted method is that the evaporation of radionuclides (Cs) can be minimized [39].

6.5 Conclusions

This chapter has discussed the application of microwaves to process different kinds of materials, which have potential usage in promising applications.

The W/Cu FGM sample has been successfully prepared by employing hybrid microwave heating method in a very short processing time. Low temperature synthesis of FGM samples has become possible because of increased diffusion kinetics owing to *non-thermal microwave effects*. Poly/nanocrystalline M-type hexaferrite has been decrystallized due to increased magnetostriction and exhibited much higher saturation magnetization. The SPIONs were synthesized as an efficient drug delivery agent. The multicomponent magnetic graphene oxide nanoparticles were developed. These nanoparticles were equipped with an adequate absorption capability for heavy metal ions and therefore can be used to improve the quality of groundwater.

The MW-assisted synthesis of non-stoichiometric tungsten oxide-doped borosilicate glasses has resulted in samples with improved characteristics. These samples exhibited reduced oxidation state as compared to conventionally prepared. The reduced oxidation is found responsible for good NIR shielding ability. In another case, the MW-assisted hydrothermal synthesis of ZnO nanoparticles with high defect concentration through accelerated crystallization is carried out under microwave irradiation. The obtained ZnO nanostructures exhibited enhanced visible light photocatalysis.

References

1. Dube, C. L., Patil, Y., Kanpara, S., Khirwadkar, S. S., & Kashyap, S. C. (2014). Microwave synthesis and mechanical characterization of functionally graded material for applications in fusion devices. *Bulletin of Materials Science, 37*(7), 1607–1611.
2. Zhou, Z. J., Yum, Y. J., & Ge, C. C. (2010). The recent progress of FGM on nuclear materials-design and fabrication of W/Cu functionally graded material high heat flux components for fusion reactor. In *Materials Science Forum* (Vol. 631, pp. 353–358). Trans Tech Publications Ltd..
3. Khirwadkar, S. S., Singh, K. P., Patil, Y., Khan, M. S., Buch, J. J. U., Patel, A., & Divakar, C. (2011). Fabrication and characterization of tungsten and graphite-based PFC for divertor target elements of ITER like tokamak application. *Fusion Engineering and Design, 86*(9-11), 1736–1740.
4. Pintsuk, G., Brünings, S. E., Döring, J. E., Linke, J., Smid, I., & Xue, L. (2003). Development of W/Cu—functionally graded materials. *Fusion Engineering and Design, 66*, 237–240.
5. Roy, R., Agrawal, D., Cheng, J., & Gedevanishvili, S. (1999). *Nature, 399*, 668.
6. Mondal, A., Upadhyaya, A., & Agrawal, D. (2010). Effect of heating mode on sintering of tungsten. *International Journal of Refractory Metals and Hard Materials, 28*(5), 597–600.
7. Mondal, A., Agrawal, D., & Upadhyaya, A. (2010). Microwave sintering of refractory metals/alloys: W, Mo, Re, W-Cu, W-Ni-Cu and W-Ni-Fe alloys. *Journal of Microwave Power and Electromagnetic Energy, 44*(1), 28–44.
8. Mondal, A., Agrawal, D., & Upadhyaya, A. (2008). Microwave heating of pure copper powder with varying particle size and porosity. *Journal of Microwave Power and Electromagnetic Energy, 43*(1), 5–10.
9. Wei, C., Cheng, J., Zhang, M., Zhou, R., Wei, B., Yu, X., & Chen, P. (2021). *Fabrication of diamond/W–Cu functionally graded material by microwave sintering*. Nuclear Engineering and Technology.
10. Sutton, W. H. (1992). *Microwave processing of ceramics-an overview* (p. 269). MRS Online Proceedings Library (OPL).

11. Wurster, S., & Pippan, R. (2009). Nanostructured metals under irradiation. *Scripta Materialia, 60*(12), 1083–1087.
12. Ibrahim Ghoneim, A. (2024). *Applications of nano-ferrites in medicine*. IntechOpen. https://doi.org/10.5772/intechopen.1003615
13. Goldman, A. (2006). *Modern ferrite technology*. Springer Science & Business Media.
14. Sharma, I., & Garg, E. (2022). A review on nanoferrites in biomedical applications. *IJBAS, 11*(2), 16.
15. Yadoji, P., Peelamedu, R., Agrawal, D., & Roy, R. (2003). Microwave sintering of Ni–Zn ferrites: comparison with conventional sintering. *Materials Science and Engineering: B, 98*(3), 269–278.
16. Ganesh, I., Johnson, R., Rao, G. V. N., Mahajan, Y. R., Madavendra, S. S., & Reddy, B. M. (2005). Microwave-assisted combustion synthesis of nanocrystalline $MgAl_2O_4$ spinel powder. *Ceramics International, 31*(1), 67–74.
17. Geetanjali Dube, C. L., Kashyap, S. C., et al. (2011). Effect of Microwave Processing on Polycrystalline Hard Barium Hexaferrite. *Journal of Superconductivity and Novel Magnetism, 24*, 567–570.
18. Sutton, W. H. (1989). Microwave processing of ceramic materials. *American Ceramic Society Bulletin, 68*, 376–386.
19. Satapathy, L. N., Ramesh, P. D., Agrawal, D. K., & Roy, R. (2005). *Materials Research Bulletin, 40*, 1871.
20. Cheng, J., Agrawal, D. K., Komaraneni, S., Mathis, M., & Roy, R. (1997). *Materials Research Innovations, 1*, 44.
21. Mohapatra, J., Liu, X., Joshi, P., & Liu, J. P. (2023). Hard and semi-hard Fe-based magnetic materials. *Journal of Alloys and Compounds, 955*, 170258.
22. Endo, T. (2001). *Synthesis of inorganic materials by 28 GHz MW radiation*. Kokushikan University.
23. Saxena, N., Dholia, N., Akkireddy, S., Singh, A., Yadav, U., & Dube, C. L. (2020). Efficient microwave synthesis, functionalisation and biocompatibility studies of SPION based potential nano-drug carriers. *Applied Nanoscience, 10*(2), 649–660.
24. Saxena, N., Agraval, H., Barick, K. C., Ray, D., Aswal, V. K., Singh, A., & Dube, C. L. (2020). Thermal and microwave synthesized SPIONs: Energy effects on the efficiency of nano drug carriers. *Materials Science and Engineering: C, 111*, 110792.
25. US Department of Health and Human Services. (1999). *Agency for Toxic Substances and Disease Registry: Toxicological profile for Lead (update) PB/99/166704*. US Department of Health and Human Services.
26. Singhal, P., Jha, S. K., Pandey, S. P., & Neogy, S. (2017). Rapid extraction of uranium from sea water using Fe3O4 and humic acid coated Fe3O4 nanoparticles. *Journal of Hazardous Materials, 335*, 152–161.
27. Dubey, A., Bhavsar, N., Pachchigar, V., Saini, M., Ranjan, M., & Dube, C. L. (2022). Microwave assisted ultrafast synthesis of graphene oxide based magnetic nano composite for environmental remediation. *Ceramics International, 48*(4), 4821–4828.
28. Wang, X., Yu, S., Jin, J., Wang, H., Alharbi, N. S., Alsaedi, A., & Wang, X. (2016). Application of graphene oxides and graphene oxide-based nanomaterials in radionuclide removal from aqueous solutions. *Science Bulletin, 61*(20), 1583–1593.
29. Dubey, A., & Dube, C. L. (2023). Microwave assisted synthesized graphene oxide nanocomposites for remediation of toxic metal ions. *Nano-Structures & Nano-Objects, 35*, 101017.
30. Mandal, A. K., & Sen, R. (2017). An overview on microwave processing of material: a special emphasis on glass melting. *Materials and Manufacturing Processes, 32*(1), 1–20.
31. Kitchen, H. J., Vallance, S. R., Kennedy, J. L., Tapia-Ruiz, N., Carassiti, L., Harrison, A., & Gregory, D. H. (2014). Modern microwave methods in solid-state inorganic materials chemistry: From fundamentals to manufacturing. *Chemical Reviews, 114*(2), 1170–1206.
32. Ross, C. P., Tincher, G. L., & Rasmussen, M. (2004). *Glass melting technology: A technical and economic assessment* (pp. 73–80). Glass Manufacturing Industry Council.

33. Vaidhyanathan, B., Ganguli, M., & Rao, K. J. (1995). Fast solid-state synthesis of metal vanadates and chalcogenides using microwave irradiation. *Materials Research Bulletin, 30*(9), 1173–1177.
34. Das, B., & Mandal, A. K. (2025). Copper doping in alkali-modified zinc borate glasses and tailoring of its oxidation state via microwave heating. *Materials Chemistry and Physics, 333*, 130346.
35. Pathak, N., Pilania, R. K., Sooraj, K. P., Ranjan, M., Sarker, D., & Dube, C. L. (2025). Microwave-assisted rapid synthesis of non-stoichiometric tungsten oxide doped borosilicate glasses for NIR shielding application. *Ceramics International, 51*, 24470.
36. Owoeye, S. S., Folorunso, D. O., Aramide, F., & Okotie, B. (2024). Microwave energy assisted fabrication and characterization of 45S5 bioglass-ceramics using bio-wastes as alternative resources for biomedical applications. *Ceramics International, 50*(8), 12746–12762.
37. Mahmood, M. A., Baruah, S., & Dutta, J. (2011). Enhanced visible light photocatalysis by manganese doping or rapid crystallization with ZnO nanoparticles. *Materials Chemistry and Physics, 130*(1-2), 531–535.
38. Li, H. X., Li, B. W., Deng, L. B., Xu, P. F., Du, Y. S., & Ouyang, S. L. (2018). Microwave-assisted processing of graded structural tailing glass-ceramics. *Journal of the European Ceramic Society, 38*(6), 2632–2638.
39. Kumar, B., Das, B., Sinha, P., Jain, U., Sengupta, P., & Mandal, A. K. (2025). Assessment of Cs volatility loss during glass melting adopting microwave and conventional heating. *Journal of Non-Crystalline Solids, 657*, 123482.

MIX
Papier aus verantwortungsvollen Quellen
Paper from responsible sources
FSC® C105338

If you have any concerns about our products,
you can contact us on
ProductSafety@springernature.com

In case Publisher is established outside the EU,
the EU authorized representative is:
Springer Nature Customer Service Center GmbH
Europaplatz 3, 69115 Heidelberg, Germany

Printed by Libri Plureos GmbH
in Hamburg, Germany